A Watched Kettle Never Boils

Contemplative Thoughts on the Nature of the Universe:
Matter, Light, Time and Gravity.

Bernard Paul Badham

'Fundamental problems require fundamental solutions.'

ISBN-13:9781482525533

CONTENTS

Introduction

What follows is a collection of thoughts and ideas on the nature of the universe, matter, light, time and gravity. This book is not to be considered as an absolute proof of the theories expounded within concerning the nature and cause of gravity, but rather an exploration of our understanding of classical and modern physics illuminated by a mixture of my own ponderings, scientific reasoning, numerous calculations, experiments and teaching experience and most importantly of all: my imagination. The ultimate aim of the physics explored in this book is to understand the true cause and nature of one of the most elusive, weakest, and yet most important forces in the universe: gravity.

The style of the book is one of questioning, tracing the traditional classical views of the nature of the physical universe and incorporating our modern quantum mechanical views of physical phenomena. This book is dedicated to my son Luke, who has listened with interest to many of my expounded theories of physics, for, without his encouragement this book would never have been written. It is also dedicated to many of my past students who have inspired me by their freshness of thought, interest and questions. The explorative dialogue is aimed primarily at students of physics and indeed any who desire a deeper and more satisfying understanding of the universe around us. The content is presented in much the same way as I have taught, with questioning, reasoning, and presentation of the known facts.

Although much proof, both theoretical and mathematical is presented in this book where necessary, the mathematical proofs and calculations are kept separate from the flow of the text and are included for those students who will find the exercises most useful in their own studies. I have deliberately avoided using references in the book as much of the physics presented is standard and modern. It is in revisiting and questioning this physics which brings us to a deeper understanding, and in doing so, by taking down the building blocks of what we know and understand and rebuilding them do we sometimes get the true picture. I have simply taken a very large jigsaw puzzle put together over many centuries by intellectual giants of physics and put them back together with logic and hopefully inspired but reasoned imagination. This is not a complete work in itself as I have not included all my writings, thoughts and experimental results, but only those which are necessary to complete the journey of discovery into the true nature of gravity.

It is my hope that you will find it a rewarding and enjoyable experience discovering anew the nature of our remarkable universe and that you will arrive at the same satisfying place as I of understanding what gravity really is.

Bernard Paul Badham

Stuck on a Rock

Humankind has been living on planet Earth for millions of years and almost all of that time our meager and mortal existence has been confined to the surface of this rocky world. It is only in recent history through the inventions of adventurous scientists and engineers have we been able to take to the air and then to finally fly to the Moon. In a sense we have been and are prisoners of our own world. It took an immense amount of research, knowledge, technology, money and chemical *power* to send men and women into space and to ultimately get the three astronauts of Apollo 11 to the Moon.

What keeps us here is of course gravity. That strange and mysterious force of nature so well described by Isaac Newton and Albert Einstein, but even with their great mathematical accomplishments we are not much wiser concerning the true nature and cause of this imprisoning force.

To lift a load against the force of gravity takes energy, and energy costs. It took the whole mass and fuel of the enormous Saturn V rocket to lift its little payload of Command Module, Service Module and Lunar Module into orbit about the Earth. It seems we are stuck in a *gravity well* and to leave this planet we have to claw our way out by any and every means possible. One very real question remains: when we are moving against gravity what exactly are we moving against? Although Newton described gravity as a force, an impulse given to a mass near another mass, Einstein declared gravity to *be* acceleration. Neither of these brilliant scientists tackled the real physical problem of what causes gravity and why and how it acts.

Gravity is described as an 'action at a distance' force, an invisible force which acts through the void of empty space, in other words it appears to act through nothing! It is no longer enough to say, as Einstein did, that mass tells space-time how to curve and space-time tells a mass how to move.

Until we understand the true nature of gravity and how to nullify its effects we will remain:

Stuck on a rock.

CHAPTER ONE
The Perception of Time

Ever since I can remember I have been fascinated by the vastness and incredible beauty of our wonderful universe and particularly one of its most elusive forces, the force of gravity. Even though I quickly learned about Newton's laws which describe accelerated motion and later Einstein's mathematical descriptions of the nature of space-time I was still left wanting. Gravity, like magnetism, was one of those mysterious invisible forces, which left me wondering how and why these forces act.

To start this incredible journey of discovery into understanding the fundamental nature of the universe we begin with time itself. We are all familiar with the concept of time, for without it how would we ever get things done? But time remains as elusive as the true nature of gravity – it is one of the hardest things as a physicist to define. I remember as a child being told by my elder sister the old adage 'a watched kettle never boils' and understanding what she meant when making a pot of tea for the family one evening, standing there in the kitchen waiting for the kettle to boil on the gas stove seemed to take forever and what was even more frustrating there was nothing I could do, or so I thought at the time, to make the kettle boil any faster. Finding the experience too much to tolerate I left the kitchen and returned to the living room to watch TV and guess what? It seems no sooner than I had made myself comfortable, that the little kettle on the stove started whistling, 'maybe she was right,' I thought. So what was going on here? Even *at the time*, I knew the *adage* or *my personal perceived experience* of the event wasn't true, what made it so real was my *perception of time*, not time itself. So then, how do we actually perceive time? Well let's start with what we all recognise as something which marks out the pace of time: a clock.

Clocks come in all shapes, types and sizes but they all essentially do the same thing, *mark the passage of time.* My *all time* favorite clock is the grandfather clock, what a wonderful piece of machinery, clock making is not just a science it is an art and if you ever open up the back of a large clock, or even a small one for that matter and take a look inside at its intricate workings of winding cogs you will agree. My first experience of a grandfather clock was when visiting an uncle of mine in his old Hereford country cottage, it was evening time and the log fire was roaring in the stone fireplace illuminating the quaint little living room with its soft flickering light and placed on the table behind me was one of those quaint old oil lamps. The soft dancing shadows being cast around the room as my parents talked were just as charming as the *tick-tock* of the tall grandfather clock set against the living room wall. I remember looking up at its ornate face; the clock stood taller than I, and watching the long brass pendulum swing hypnotically to and fro, the sound and sight of the brass pendulum gracefully beating out the passage of time was magical.

Tick – tock, tick – tock, tick – tock

The experience of this audible beat of time, where each *tick* and *tock* was governed by the swing of the pendulum was somehow wonderfully soothing and relaxing. If you ever have trouble sleeping my advice is to buy a grandfather clock and listen to its rhythmic *tick-tock* while relaxing in a comfortable chair, preferably with the heat of a log fire in front of you and with no other distractions except the flicker of the flames.

Tick – tock, tick – tock, tick – tock

Listening to the clock and watching the pendulum swinging to and fro gives the impression that time marches on evenly and uninterrupted, and so by this premise my boiling kettle would have taken the same time to boil whether I was watching it or not.

Tick – tock, tick – tock, tick – tock

So what exactly governs our perception of time? First there is our *psychological perception of time*, which is relative of course to what we are doing when we are experiencing the passage of time, 'time flies' when we are enjoying ourselves, but seems to go by slowly when for instance we are waiting in a queue. Secondly, there is the *biochemical awareness of time*: On the scientific biological level the speed of the chemical reactions in our brain, our thought processes, govern our perception of time.

Chemical reactions involve interactions of electromagnetic fields between atoms and the speed of these interacting fields is governed by the speed of light, which is the constant (c) equal to 3×10^8 m/s, an incredible speed of over one hundred thousand kilometres and hour!

It is the constant ticking away of these reactions, at the speed of light, that give us a constant biological perception of the passage of time. It is interesting to note here that if we could change the speed of light then the speed of our thinking processes would also change, if increased then we would think faster, but so would all the physical processes in the world around us work faster, so we would not notice any change at all!

The kettle would actually boil faster, but because we are thinking faster it would seem to take just as long!

Lastly, there is what we call *absolute time*, the thing clocks measure. So what exactly is absolute time? There are various answers to this fundamental question: 'Well time is the *time* it takes for the Earth to spin around once which takes 24 hours!' 'Time is something governed by clocks!' 'Time is the interval between one regular event and another!'

Interestingly enough is the fact that time *can* be measured and since it can be measured, this makes it a fundamental physical quantity. One revolution of the Earth about its axis with reference to the Sun is what we call one solar day, 24 hours. This time duration of one solar day was divided up by the ancient Egyptians into 12 hours of day and 12 hours of night. The passing of day time hours were marked by the movements of the sun and water clocks, whereas the night time hours were marked by the movements of the stars. The priests of ancient Egypt would call out the hours from the temple rooftops each time a specific star crossed the line of sight of a fixed apparatus.

If the day is measured with respect to the 'fixed stars' we get a sidereal day, which is actually shorter than the solar day: The duration of a sidereal day is 23 hours, 56 minutes and 04.0905 seconds with respect to the Mean Solar Time of Greenwich Meridian time. The sidereal day is shorter because the Earth *moves* in its orbit around the sun as it revolves about its axis. As we know today one revolution of the Earth in its orbit about the sun is what we class as one year, 365.2422 days or 365 days, 5 hours, 48 minutes and 45 seconds.

Absolute time is a very difficult thing to define, ancient philosophers and scientists grappled with the subject of time: The ancient Greek philosopher, Heraclitus said: 'Everything flows and nothing abides; everything gives way and nothing stays fixed. You cannot step twice into the same river, for other waters and yet others, go fluxing on.

Time is a child, moving counters in a game; the royal power is a child's.' Another ancient Greek philosopher, Antiphon said: 'Time is not a reality, but a concept or a measure.' St Augustine confessed: 'What then is time? If no one asks me, I know: if I wish to explain it to one that *asketh*, I know not.'

A dictionary definition gives time as: 'A non-spatial continuum in which events occur in apparently irreversible succession from the past through the present to the future.' Or 'An interval separating two points on this continuum; a duration.' An encyclopaedic definition gives it as: 'Time is part of the fundamental structure of the universe, a dimension in which events occur in sequence, and time itself is something that can be measured.'

All of these definitions are of course unsatisfactory to varying degrees, so what exactly is time? I once asked my youngest son this question and he gave the best answer I have ever heard: 'Dad, time is change!' a simple but very real answer, for without time there can be no change. So a time interval is an interval of change. So what governs the rate at which things change? This we know and have already stated: *the speed of light!*

The speed of light governs change, therefore governs time

Things change at the speed of light: all fundamental forces, electromagnetic and gravitational, act at the speed of light. So in this sense: time *is* the speed of light! Or more exactly:

Time is fundamental change at the speed of light

But what is even more interesting is that according to Einstein, relatively speaking, time is not constant, only the speed of light in a vacuum is constant and that: speed and gravity affects time, this is what he calls time dilation, at high near light velocities or in a strong gravitational field time slows down relative to a stationary observer or someone in a weak gravitational field, for each observer in their own frame of reference time runs at its normal speed. This constancy of the speed of light and time for each observer in their own frame of reference brings order and harmony to the universe.

Albert Einstein was a giant of logic, reasoning and imagination, there has never since been a scientist like him. He was a truly intellectual giant, for 100 years later, his theories still stand. Einstein's time dilation, due to the effects of speed and gravity, *are* very real.

Navigation satellites in orbit travelling at high velocities and where gravity is weaker than on Earth have to compensate for these time differences. So why does speed and gravity affect time? These are two very important fundamental questions; let's answer the first why gravity affects time:

In a strong gravitational field the <u>speed of light is reduced</u> and since the speed of light governs time, gravitational fields affects time.

Why speed has an effect on time will be answered later when we have looked at the physical nature of space though which objects move.

These time dilation affects are very small, almost immeasurable, but they *have* been measured using modern atomic clocks. What would happen to our grandfather clock in different strength gravitational fields? There are two affects:

1. The large and noticeable affect due to the change in the gravitational acceleration of free fall.

2. The small but measurable affect of time dilation.

In a strong gravitational field, the free fall acceleration of a swinging pendulum of a grandfather clock increases, so the clock would visibly tick faster, on the Moon for instance where gravity is one sixth that of Earth's a pendulum clock would tick slower (one sixth slower). This difference we would notice.

The time dilation due to changes in gravity we would only notice between different frames of reference and in any way if we could observe this affect it would be very small. What about our boiling kettle in different gravitational fields? Well on the surface of the Earth where gravity is strong and time runs slower, the kettle would take longer to boil, but we would not notice this because light speed is less and therefore our brain's thought processes, the chemical reactions, are running slower by an equal amount. If we take the kettle to the top of a mountain where gravity is less and time runs faster, the kettle would boil quicker, but again we would not notice this in our own frame of reference. Is there any way we can make the kettle boil quicker? Yes, but not in our own frame of reference!

One solution to this problem is to place the kettle inside a magical box were gravity is decreased artificially to zero, we would then observe the kettle boiling quicker, but because the time difference between zero gravity and Earth's gravity is minute, it would still seem to take just as long, we need a bigger gravitational difference between our frame of reference and the kettle inside the box. The problem is we would probably have to lie down while we are waiting for the kettle to boil because if we were in an artificially increased gravitational field we would weigh more!

Creating an artificial gravitational field is a holy grail for physics. If we could control gravity then we could travel through space and time. Yes, we are talking about time machines and spacecraft travelling at warp speeds, the speed of light and more. At last we may be able to travel to the stars and travel through time!

CHAPTER TWO
The Nature of Light

Our journey into the nature of the universe will take us down many roads of classical and modern physics, some seeming irrelevant and off the track, but every fact of physics we examine will be essential in our eventual understanding of why a mass produces a gravitational field - gravity's causal agent. In other words: How matter warps space-time.

If we can understand how gravity works, then we may be on the road to creating artificially our own gravity field.

Whoever controls gravity can control space and time!

This is a fairly mind blowing statement, absurd even, a pipe dream perhaps? But whoever makes this breakthrough in the world of physics will go down in history. It will be the next quantum leap for humankind, for with the knowledge of how to warp space-time is the ability to achieve one of mankind's ultimate goals - to travel to the stars, anywhere in space and possibly time.

There are a few physicists undertaking theoretical research into the holy grail of physics - the creation of a gravitational field. Some say it is impossible, while others have claimed limited success. A few years ago the American government announced that they were about to spend millions of dollars on a gravitational field project. Most of the theories used in this book are based on real physics, but the world of physics is changing, what was once thought impossible now takes on the guise of the probable. This is an adventure into the universe of space and time, which, according to Einstein are variable, not constant. The one thing in this universe he said which remains constant is the speed of light (c) in a vacuum: In metric units, c (in a vacuum) is exactly:

$$299,792,4588 \text{ metres per second}$$

In more familiar units it is a phenomenal: 1,079,252,848.8 km/h, over a thousand million kilometres an hour. Travelling at this speed we could whiz around the whole circumference of the Earth 7.49 times a second! Although this gives us the impression that the speed of light is instantaneous here on Earth over terrestrial distances, when it comes to the size of the universe light speed is pretty slow. For instance it takes 1.28 seconds for light to reach the moon from Earth, this delay was significant for the communication of Houston ground control and Apollo 8 when it became the first spacecraft to orbit the Moon: For every question, Houston had to wait nearly 3 seconds for the answer to arrive, even when the astronauts replied immediately.

The following is a list of flight times for light from the sun reaching astronomical objects:

Earth	8.32 minutes
Mars	12.7 minutes
Pluto	5.48 hours

Nearest star: Proxima Centauri: 4.2 years!

This shows us that interstellar space travel is only possible if we could travel at near light speeds or even greater if that were possible, what we call *superluminal speeds*. To travel across our own galaxy, the Milky Way, at the speed of light would take 100,000 years and to get to the nearest galaxy, to ours, Andromeda, would take 2.9 million years.

The strange thing is that through a transparent or translucent material medium, like glass or air, light appears to have a different speed than in a vacuum, this causes its direction of propagation to change; the ratio of c in the medium to the observed velocity in a vacuum is called the *refractive index* of the medium. This travelling at different speeds through different media is called *refraction*, let us look at this property of refraction of light in more detail as later it will become important in our understanding of space-time dilation.

Refraction is the change in direction of a wave due to a change in its speed. This is seen when a wave passes from one medium to another. Refraction of light is the most commonly seen example, but any type of wave can refract when it interacts with a medium, for example when sound waves travel through materials of different density. At the boundary between different media, the wave's velocity is altered, and as a result of an uneven change in speed across the wave front it changes direction. Analogy:

Imagine yourself running past a lamp post and just as you pass it you reach out your right hand to try and catch it, the result is reduced pace and a change of direction — you swing to the right.

In refraction there is a wavelength increase if the medium is less dense and a decrease if the medium is denser, but in both cases the wave's frequency remains constant. For example, a light ray will refract as it enters and leaves glass, the understanding of this concept led to the invention of lenses and the refracting telescope. Refraction is also responsible for rainbows and for the splitting of white light into a rainbow-spectrum as it passes through a glass prism. Glass has a higher refractive index than air and the different colours of light, of different frequencies, travel at different speeds (dispersion), causing them to be refracted at different angles, so that you can see them. Violet light slows down in glass more than red light, this bigger change in its speed causes violet light to bend (refract more) than red light.

The speed of light in air is only slightly less than c, in a denser medium, such as water and glass, light can slow much more, to fractions such as $3/4$ and $2/3$ of c. respectively. Through diamond, light is much slower — only about 124,000 kilometres per second, less than $\frac{1}{2}$ of c.

Why then is the speed of light in a medium such as glass much less than the speed of light in a vacuum?

The reason for this slowing down of light as it enters a denser medium is fundamental in our understanding of gravitational fields.

When light enters a substance such as glass from a vacuum, travelling at c, the individual particles of light (photons) interact with the glass atoms. The atoms act like antennae and absorb the photons and then after a specific time interval (Δt) they transmit them. This time delay gives light an apparent reduction in speed, *in between* the atoms there is still the vacuum of space where the speed of light is constant (c). When the light photons leave the glass medium they continue at high speed, at c.

Dispersion of Red and Violet Light

Let us look again at the dispersion of red and violet light in water from air and see if we can explain the differences in speed. We first calculate their respective speeds in water using known values of refractive index (n).

$$n_{air} = 1.000 \quad n_{red} = 1.331 \quad n_{violet} = 1.344$$

1. Speed of red light (v_r) in water

$$= c/n = 2.998 \times 10^8 ms^{-1}/1.331$$

Velocity Red Light $v_r = 2.252 \times 10^8 m/s$

2. Speed of violet light (v_v) in water

$$= c/n = 2.998 \times 10^8 ms^{-1}/1.344$$

Velocity Violet Light $v_v = 2.231 \times 10^8 m/s$

It can be seen from the calculation that violet light has a slightly slower speed in water than red light and therefore there is a bigger change in the speed for violet light entering water from air, this bigger change in speed causes more refraction – a bigger direction change.

But why does violet light travel slower through water than red light? Surely the answer to this question must be related to why light slows down when travelling through a dense medium and this is related to the photon absorption time (Δt) by the atoms of the medium. Since violet light slows down more, then the photonic absorption time for violet light must be greater than the absorption time for red light. Assuming this is true then we must ask: why is the absorption time greater? Violet light has a shorter wavelength (400nm) compared to red (700nm), using the wave speed formula we can calculate their respective frequencies and energies:

Red light frequency f_r

$$= c/\lambda = 2.998 \times 10^8 ms^{-1}/700 \times 10^{-9}m$$

$$= 4.283 \times 10^{14} s^{-1}$$

Violet light frequency f_v

$$= c/\lambda = 2.998 \times 10^8 ms^{-1}/400 \times 10^{-9}m$$

$$= 7.495 \times 10^{14} s^{-1}$$

Now we can calculate the energy of each photon using the Photon Energy equation $E = hf$:

Where h is Planck's constant (h) = 6.626×10^{-34}Js:

Red photon energy (E)

$= hf = 6.626 \times 10^{-34}$Js $\times 4.283 \times 10^{14}$s^{-1}

$= 2.838 \times 10^{-19}$J

Violet photon energy (E)

$= hf = 6.626 \times 10^{-34}$Js $\times 7.495 \times 10^{14}$s^{-1}

$= 4.966 \times 10^{-19}$J

We can see that since violet light has a higher frequency of oscillation than red light, each photon of violet light carries more energy than red light photons. It seems therefore: the more energy a photon carries the greater the absorption time and therefore the slower the speed through the medium – this answers the question why violet light is refracted more than red light. Now one question remains: why does a photon carrying more energy have a greater absorption time? When an atom absorbs photons of light this shifts the electrons around the atoms to higher orbital energy levels – this is called an electron energy transition. In this higher energy state the electron orbit is unstable and within nanoseconds the electron will drop back down to its lowest energy state (ground state) and emit the photon on its way again. The time for this process is called the transition time (Δt). It seems logical to assume the higher the energy of the photon the higher the energy jump and therefore the longer the transition time – think of throwing a ball higher into the air, the more energy you give it, the higher it will go and the longer it will take to come back down. In this case the higher the photon energy the slower the propagation speeds through the medium.

Classically, considering electromagnetic radiation to be like a wave, the charges of each atom (primarily the electrons) interfere with the electric and magnetic fields of the radiation, slowing its progress. [The full quantum-mechanical explanation is essentially the same, but has to cope with the discrete particle nature of light: the Electric fields in the atoms create phonons in the media, and the photons mix with the phonons. The resulting mixture, called a polariton, travels with a speed different from light.]

Here is an important principle to remember:

Light speed c in a vacuum is a constant, but in different media its speed can change.

But what exactly is light? Understanding the nature of light is fundamental for our understanding of the nature of matter, energy, gravity and space-time. To help us understand any phenomena we look at its measurable physical properties and effects. Classical physics states: that light is an electro-magnetic wave, which is visible to the naked eye (what we call visible light), but light in a technical or scientific context, is electromagnetic radiation of *any* wavelength. Red light has a longer wavelength (lower frequency) than blue light: The wavelength of red light is around 650 nanometres i.e. 650×10^{-9}m. The wavelength of blue light is 450 nm. The visible spectrum of light includes the familiar colours of the rainbow.

The electromagnetic spectrum encompasses all electromagnetic waves including waves which are invisible to the naked eye, these include radio waves, microwaves etc. All of these waves have the same basic properties of light; they can be reflected, and refracted. The three basic dimensions of light (all electromagnetic radiation) are: Intensity (I) measured in Watts per metre squared, or alternatively amplitude (a), which is related to the perception of brightness of the light and the height of the wave. Frequency (f) measured in wave cycles per second (s^{-1} or Hertz, Hz). Wavelength (λ) measured in metres, perceived by humans as the colour of the light, and Polarization (angle of vibration), which is only weakly perceptible by humans under ordinary circumstances.

The wave speed (v) can be calculated by using the equation:

$$\text{Wave speed} = \text{frequency} \times \text{wavelength}$$

$$V = f\lambda$$

When we are talking about light and other electromagnetic radiation which travels at the speed of light we use the symbol (c), so:

$$c = f\lambda$$

Due to the wave-particle duality nature of light, light simultaneously exhibits properties of both waves and particles. The elementary particle that defines light is the photon. A photon is a packet (quantum) of electromagnetic energy. The electromagnetic energy (E) of a photon at a particular wavelength λ (in a vacuum) and its associated frequency (f) can be calculated:

Photon energy = Planck's constant x frequency

$$E = hf$$

The value of Planck's constant is: 6.626 0693 x 10^{-34} Js.

Question: What is the electromagnetic energy of a photon of red light of wavelength 650nm?

$$E = hf$$

$$E = 6.6261 \text{ x } 10^{-34} \text{Js x } 650 \text{ x } 10^{-9}\text{m}$$

Answer: The electromagnetic energy of a red photon of light = 4.04 x 10^{-40} joules

Planck's constant (h) is a fundamental constant of the universe, it determines in this case the energy of an oscillating electromagnetic wave and as we shall see later it governs many other phenomena including the smallest size attainable: 10^{-35}m, what we call Planck length: It is interesting to note here that Planck's constant (h), the speed of light (c) and Newton's gravitational constant (G) are related.

Question: How many photons does a 100 Watt light bulb radiate per second?

A 100 Watt red light bulb radiates 100 Joules of photonic energy per second.

Number of photons per second = Energy per second/energy per photon

$$= 100\text{J}/4.07 \text{ x } 10^{-40}\text{J} = 2.46 \text{ x } 10^{41}$$

photons per second!

Answer: A 100 watt red light bulb radiates about 25 billion, billion, billion, billion, billion photons per second!

Note: In physics there are theoretical limits on size etc, one such theoretical limit is called Planck length. The Plank Length limits the smallest theoretical particles which can exist. Planck Particles which are 10^{20} times smaller than a proton have a mass which is about 13 x 10^{18} times heavier than the mass of a proton:

In quantum physics the more massive a particle

the smaller its size!

Planck length, denoted by Lp, is the unit of length about 10^{20} times smaller than the size of a proton in an atomic nucleus. The Planck length is a natural unit because it can be defined from three fundamental physical constants: the speed of light (c), Planck's constant (h) and the gravitational constant (G)

Question: how can Planck length be calculated using the constants of light speed and gravity?

Answer:

$$L_p = \text{square root } (hG/2\pi c^3)$$

$$= \text{sqrt } (6.626 \times 10^{-34}\text{Js} \times 6.673 \times 10^{-11}/2\pi \times 2.998 \times 10^{8}\text{m/s})$$

$$L_p = 1.616 \times 10^{-35}\text{m}$$

It is interesting to note here that the limit of Planck length is based on the gravitational constant G, which cannot be derived from other constants, G can only be determined by physical measurement.

Let us investigate the electro-magnetic nature of light, because this is fundamental in understanding the true nature of matter, energy, space-time and gravity. We will start with Einstein's mass-energy equivalence:

This simply states that mass and energy are equivalent, mass can be formed from energy and vice versa – matter and energy are interchangeable.

A simple example: a photosynthetic plant locks up light energy which has been radiated from the sun over the vacuum of 'empty' space. The green photosynthetic pigment, chlorophyll, traps this electromagnetic light energy into the energy rich carbohydrates, glucose and starch which we use for food. Photosynthesis traps electromagnetic light energy into mass. This is an example of an energy-mass conversion; any object which absorbs light energy increases its mass.

Since Einstein we understand that:

Matter can be transformed into electromagnetic wave energy

Electromagnetic wave energy can be transformed into matter!

Einstein's mass-energy equivalence equation:

$$E = m c^2$$

This equation tells us how much electromagnetic energy we can get from a mass:

Energy = mass x speed of light squared

Multiplying the mass by the speed of light squared means that it takes an awful lot of energy to make one kilogram of mass or in one kilogram of mass there is an immense amount of energy. In a one kilogram mass there is: 90,000,000,000,000,000 Joules of electromagnetic energy:

Electro-magnetic energy trapped in 1 kg of matter:

$$E = mc^2 = 1 \text{ kg} \times (2.998 \times 10^8 \text{m/s})^2$$

$$\text{Energy} = 8.998 \times 10^{16} \text{ Joules}$$

This principle is proved every time you strike a match - a tiny amount of matter is converted into electromagnetic energy - heat and light. The heat is in the form of electromagnetic infra-red waves and the kinetic (movement) energy of the gas particles. The same thing happens whenever we burn anything.

Mass-energy conversions happen in all chemical reactions.

It's what keeps us alive - our body 'burns' food chemically to release the mass-energy. A more dramatic example of mass-energy conversion is an atomic bomb. Nuclear power plants do the same thing - radioactive uranium is split (fission) into two simpler elements which are less massive together than the original uranium. This missing mass is converted into pure energy, mostly in the form of moving (kinetic) heat energy. The Sun and all other stars are the most efficient machines at converting matter into energy. They do this by the process of nuclear fusion in the core of the star. Here 4 hydrogen atoms are fused into 1 helium atom. The helium atom is less massive than the 4 hydrogen atoms; hence the mass difference gets converted into pure electromagnetic energy (gamma rays). On a smaller everyday scale the mass-energy conversions are tiny:

1. Heating a 1kg copper pot from 0°C to 100°C takes around 40kJ of energy. This is equivalent to a mass increase of 10^{-13}kg, a tiny increase in mass, but still present and real.

2. Hitting a tennis ball from a velocity of 0m/s to 50m/s gives it about 125 joules of kinetic energy, this is equivalent to a mass increase of 1.4×10^{-15}kg.

3. Energy absorbed per second by the Earth from the Sun is a staggering 1.74×10^{17}J, this is equivalent to a mass increase in the Earth of 1.93kg per second. Don't worry the Earth is not getting bigger by 60.82 million kg a year, because most of this mass-energy is re-radiated back out into space.

The sun in its core converts mass to energy at a phenomenal rate: 4.26 million metric tons per second! This energy is released in the form of high energy neutrinos and high frequency gamma rays. The neutrinos, which neutrally charged, almost massless particles, pass through solid matter and leave the sun effortlessly. About 65 billion neutrinos from the sun pass through 1 square centimetre of the Earth's surface every second! Very few of these will interact with the Earth, almost all pass straight through as if the Earth was not there, matter is transparent to neutrino radiation. Apparently it takes on average 1 million years for a gamma photon made in the core to reach the sun's surface, this is because it is being continually absorbed and re-transmitted by matter particles in random directions. By the time this gamma radiation emerges from the sun its energy has been reduced into longer wavelength radiation: the electromagnetic spectrum, including visible light. It's an amazing thing to realize that a lot the sun's light we see today was made in its core a million years ago.

Electromagnetism

Let's get back to the electromagnetic nature of light, to do this we need to understand electro-magnetism. We have all played with magnets and built simple electrical circuits, so we have an experience and notion of electricity and magnetism.

Magnetism and Magnetic Fields

Let's talk about magnetism first. We understand that magnets have magnetic poles, a north-pole (N) and a south pole (S) and that two like poles (N-N) repel each other and so do two south poles (S-S). Opposite poles (N-S) attract. Magnetism is a force in that it can attract or repel, *unlike gravity which can only attract*. This is one of the odd things about gravity, it is always an attractive force – herein lies a clue to its nature. We are all familiar with the school experiment at sprinkling iron filings a piece on paper placed over a bar magnet. The iron filings form a pattern around the magnet which appears as loops extending from one pole to the other, this demonstrates the magnetic field around the magnet, think of it as a magnetic force field. The magnetic field has strength and direction, the closeness of the lines of force in a field diagram shows the strength of the field and the arrows show the direction. The magnetic field is stronger near the poles and the direction of the field flux (flow) from the North Pole (N) to the South Pole (S).

The direction of the magnetic field is determined by placing a compass in the field and seeing which way it points. The strength of the magnetic field is a measure of the force it exerts on another magnet like a compass needle. If we place two magnets together, a north pole (N) and a south pole (S) then the field lines line up directly between the poles. With two magnets in repulsion the field lines push away from each other and produce a neutral point in the centre space between the magnetic poles, here the opposite direction of the fields cancel each other out. The fields are still there, but a magnetic particle at this point is equally pulled in both polar directions and therefore does not move. The point of all this is to show that a magnetic field (abbreviated as B) has direction.

An electromagnetic wave is made up of two oscillating fields, an electric field (E) and a magnetic field (B). The magnetic field (B) changes direction every half wavelength i.e. twice during one oscillation of the wave. The oscillation of the wave is its electromagnetic energy.

The more it oscillates per second (frequency) the more electromagnetic energy the wave has.

We saw *this in the equation* $E = hf$

Increase the frequency and the energy of the wave increases. Think about waving a flag, to make it oscillate quicker you have to put more effort (energy) into it. What about the electric field? Magnetism and electricity go hand in hand, we can't have one without the other, so let's talk about electric fields.

Electricity and Electric Fields

Let us start this by reminding ourselves what electricity is. If we connect a lamp to a battery, electricity (electric current) travels through the wires and the lamp - the electrical energy heats up the lamp until it is white hot. The small filament in the lamp gives off heat and light energy.

Question: what exactly is travelling through the wires?

Answer: negatively charged electrons.

But what are electrons? Electrons are the negatively charged particles which 'orbit' the positively charged nucleus of an atom. The nucleus contains positive protons and electrically neutral neutrons. The atom is normally electrically neutral with the same number of negative and positive charged particles:

Number of electrons = Number of protons

In metal conductors such as copper many of the outer electrons of the atom are free to move, we call these 'free electrons.' It is these free electrons which are pushed around the circuit by the battery voltage. The more voltage the more the electrons are pushed. The negative terminal of the battery is a supply of free electrons – here the electrons are repelled from each other - because they have the same charge (-). This is a bit like two magnetic poles repelling each other. The free electrons trying to get away from each other travel through the wire - they jump from atom to atom in the wire. At the same time as being pushed through the wire by their repulsion of each other, they are also attracted to the positive (+) terminal of the battery which is deficient of electrons. So the battery supplies a push-pull on the electrons moving them through the circuit from (-) to (+). The electrons drift through the metal lattice, colliding with the atoms and giving up some of their energy as electromagnetic heat and light radiation.

Charged particles like magnets have a force field around them, electric field lines in this case. The direction of the electric field (E) is the direction a positively charged particle would move if placed in the field, repelled from the positive and attracted to the negative.

If we replace the lamp with two metal plates with an air gap in between them, the electrons are pushed by the (-) terminal and build up on one of the plates - the plate becomes negatively (-) charged. At the same time the (+) terminal of the battery attracts away electrons from the other plate - leaving it with a positive (+) charge. The end result is one of the plates is negatively (-) charged and the other plate is positively (+) charged. The electrons on the negative plate are attracted to the positive plate but cannot flux across the air gap because air is an insulator - it would take a very high voltage (strong attractive force) for the electrons to force their way through the atoms of the air insulator, if the voltage is high enough we would see them jump the gap as an electric spark!

This type of device is called a capacitor - it stores charge. What is interesting is that between the plates (+) and (-) plates we have an electric field (E)

The direction of the field is from positive (+) to negative (-).

The strength of this electric field depends on the voltage of the battery, the dimensions of the plates and the type and size of gap between them. The *electric field (E) between the plates can exert a force on any charged particle* placed between the plates. A positively charged (+) particle placed between the plates will move (accelerate) towards the negative plate and conversely, a negatively charged (-) particle with be attracted towards the positive plate.

In an electromagnetic wave the electric field is oscillating in tune with the magnetic field - they are inseparable and linked. The electric field generates the magnetic field and vice versa. Another thing to note about the electric (E) and magnetic (B) fields is that they are always at right angles to each other. E and B are also at right angles to the direction the wave travels - this is what we call a transverse wave.

Question: how is electromagnetic radiation produced?

Answer: one important way is by the acceleration of charged electrons.

When an electron accelerates, changes direction or speed, it radiates electromagnetic wave energy. This is how radio waves were first produced. In, 1864, James Clerk Maxwell predicted the existence of radio waves by means of mathematical model. Twenty four years later, in 1888, bolstered by Maxwell's theory, Heinrich Hertz first succeeded in showing experimental evidence of radio waves by his spark-gap radio transmitter. This experiment stimulated Marchese Guglielmo Marconi, who first achieved signal transmission by means of radio waves over 10 m in 1895 and over the Atlantic Ocean in 1901. It was Reginald Fessenden who first succeeded in transmitting continuous wave (CW) for voice telecommunication. These very early transmitters included a battery power supply, a high voltage induction coil (transformer) with a buzzer-type interrupter in the transformer's primary circuit, a spark gap connected across the secondary coil, and an UHF dipole antenna connected across the spark gap. The transmitted frequency was around 400 million Hertz (microwaves).

The capacitor charges up to a high voltage and is discharged across the rotary gap – a spark, the discharge current excites the antenna and charge (electrons) oscillate back and forth (accelerating) in the antenna at the antenna's natural frequency. Because the electrons are in a constant state of acceleration they radiate electromagnetic radiation. The electromagnetic wave oscillates at the same frequency (around 3×10^9 Hz) as the current in the antenna. The antenna is made up of two rod conductors, an electric dipole. Each half of the dipole is of opposite charge to the other and with a high voltage difference and alternates positive (+) to negative (-). During the A.C. voltage cycle the potential difference between the two halves is high enough for a spark. The electric field around the dipole is doughnut shaped and in the same direction as the vertical dipole. As the field oscillates, changes direction, its electromagnetic energy disperses at the speed of light forming an electromagnetic wave. We must remember of course that there is also an oscillating magnetic field at right angles to this oscillating electric field.

Summary

The Seven Fundamental Properties of Electromagnetic Waves:

1. All electromagnetic waves travel at the speed of light (c)

2. Electromagnetic waves can travel through the vacuum of space

3. The speed of an electromagnetic wave in a vacuum is a universal constant (c)

4. The wave oscillates with electro-magnetic energy at a frequency (f)

5. The electromagnetic energy (E) is directly proportional to the frequency (f)

6. The wave has an oscillating electric (E) and magnetic field (B).

7. The electric and magnetic fields oscillate at right angles to each other and the direction of the wave travel.

Mass-Energy Equivalence: *By Einstein's $E = mc^2$:*

> *Matter can be transformed into electromagnetic energy*
>
> *Electromagnetic energy can be transformed into matter*

The fact that matter (mass) and energy are interchangeable is an important milestone, for until we can understand what matter is truly made of we cannot understand why a mass produces a gravitational field.

CHAPTER 3
A Weighty Problem

Isaac Newton was a giant in our understanding of the physical universe. He laid down the fundamental laws of motion and forces, particularly the force of gravity as we understand it today.

It is strange to think that all mass (substance) is condensed electromagnetic wave energy, but daily we see and experience these mass energy conversions: Whenever we light a fire the bound up mass-energy stored in the fuel is released as heat and light (electromagnetic) energy. Burning, coal, oil or gas which are fossil fuels, releases the mass-energy stored by living organisms millions of years ago. These organisms, primarily plants, trapped the sun's electromagnetic light energy while they lived. Now we burn their fossils to release their stored mass-energy. A motor car burns petroleum (made from oil) to release stored mass-energy. The energy of the chemical combustion of the fuel in the presence of oxygen is released as heat energy.

This type of energy comes from the breaking and making of chemical bonds between the atoms. The mass differences between the products and the reactants due to the breaking and forming of new chemical bonds, is released as energy. The energy is released in the form of kinetic (movement) energy of the molecules which we see as heat. This is just like releasing a squashed spring where the stored strain energy is released causing the spring to fly off at velocity. These excited atoms release this energy in the form of Infra Red radiation, in fact all matter particles above a temperature of absolute zero (0K, -273 Celsius), radiate heat waves, the more excited the atoms (hotter), the higher the frequency/energy of the radiation which is emitted.

So how does condensed electromagnetic wave energy which makes up mass generate a gravitational field? Let's start with Isaac Newton's views on the subject. In 1679, Newton returned to his earlier work on mechanics, i.e., gravitation and its effect on the orbits of planets. He published his results in De Motu Corporum (1684). This contained the beginnings of the laws of motion that would inform the Principia: *The Philosophiae Naturalis Principia Mathematica* (now known as the Principia) was published on 5 July 1687). In this work Newton stated the three universal laws of motion that were not to be improved upon for more than two hundred years. He used the Latin word gravitas (weight) for the force that would become known as gravity, and defined the law of universal gravitation.

Note: Newton's laws of motion, as you shall see later in the book, prove invaluable to understanding the nature of our universe, space, matter and time and his laws give us an insight into the physical nature of gravity and the mechanism by which it works. In every scientific theory there is a cornerstone of truth which holds the whole together, it took many months of searching while writing this book to find a cornerstone of truth on which the quantum gravitational theory expressed in the later chapters holds true, and to my surprise and delight, it was Newton's laws of motion, specifically his second law, written some four hundred years earlier which became the cornerstone.

Newton's law of universal gravitation

Newton's law of universal gravitation states the following:

Every mass object in the Universe attracts every other mass object with a force directed along the line of centres of mass for the objects.

The force between two masses is proportional to the product of their masses and inversely proportional to the square of the separation between the centres of mass of the two objects.

Given that the force is along the line through the two masses, the law can be stated symbolically as follows.

$$F = - G \, m_1 m_2 / r^2$$

Where:

F is the magnitude of the (attractive) gravitational force between two objects,

G is the universal gravitational constant: $G = 6.673 \times 10^{-11} \, Nm^2kg^{-2}$,

m_1 and m_2 are the masses of first and second object,

and r is the distance between the objects.

It can be seen that this force F is always negative, and this means that it is always an attractive force unlike other forces in the universe, electrostatic and magnetic, which can be both attractive and repulsive. The gravitational attraction between the masses is proportional to the product of the masses of each object, but there is an inverse square relationship with the measured distance between the centres of the masses.

If the masses m_1 and m_2 are doubled the force increases: $2 \times 2 = 4$ times

If the distance between the masses is doubled the force decreases:

$1/2 \times 1/2 = 1/4$ ie deceases 4 times

Gravity is classed as the weakest of the four known forces but it acts to infinity. The weight of an object on the surface of a planet is determined by the mass of the object and its location in the planet's gravitational field:

Weight of an object $= mg$

Where 'm' is the mass of the object and 'g' is the gravitational field strength on the surface of the planet:

Gravitational field strength $g = GM/r^2$

Question: what is the magnitude of the gravitational field strength 'g' at the surface of the Earth?

Answer: If M is the mass of the planet Earth, then, for the surface of the Earth the gravitational field strength 'g' is:

Weight $= m \times GM/r^2$

$= 6.67259 \times 10^{-11} Nm^2 Kg^{-2} \times 5.976 \times 10^{24} kg / (6.378 \times 10^6 m)^2$

$g = 9.802$ N/kg (newtons per kilogram)

Question: what is the magnitude of acceleration for a 1kg mass in free fall?

Answer: by Newton's Second Law of Motion:

Acceleration $=$ force/mass

$a = F/m = 9.802/1$

$a = g = 9.802$ m/s^2

Acceleration of free fall $g = 9.802$ m/s^2

What this means is that near the Earth's surface a free falling mass, with no air friction, will increase its velocity by 9.802 m/s every second. We refer to this acceleration as 'g' - gravitational acceleration. This gravitational acceleration 'g' of 9.802 m/s^2 is a constant for all masses in free fall no matter what the mass, for as we double the mass of an object, the gravitational force on it from the Earth doubles, so although the object is twice as difficult to accelerate, there is double the force, producing the same acceleration for all masses in free fall.

In free fall all objects fall at the same rate

To prove 'g' is constant for different masses try this little experiment:

1. Fall with air resistance: drop a coin and a small piece of paper together; the coin will hit the ground first because the air drag affects the paper more.

2. Free fall: Now place the small piece of paper on top of the coin to shield it from the air resistance and then drop the coin. Both hit the ground together!

This concept that light objects fall at the same rate as heavy ones was first demonstrated by Galileo Galilei (1564-1642): He is said to have dropped balls of different masses from the Leaning Tower of Pisa to demonstrate that their time of descent was independent of their mass (excluding the limited effect of air resistance). This was contrary to what Aristotle had taught: that heavy objects fall faster than lighter ones, in direct proportion to weight.

The Importance of Gravity

Gravity is a force which acts between any two masses, *anything that has mass (substance) has gravity.*

Gravity acts between all masses in the universe

It is the gravitational force which keeps a cup on a table, causes an object to fall, keeps the Moon and planets in orbit, and keeps the stars together in our galaxy the Milky Way. It is the force of gravity which powers the Sun and any other star.

Question: how does gravity power stars?

Answer: a star such as our sun tries to collapse under its own gravity; this creates enormous pressures and temperatures in the core of the Sun making it hot enough in the core to drive nuclear fusion. The surface temperature of the sun is 5,578K and in the core it reaches a staggering 15.7 million K.

These extreme conditions keep the processes of nuclear fusion of hydrogen into helium running - the result is the release of an enormous amount of energy.

Gravity is the force which drives our universe!

Newton's Weighty Problem

In Newton's world gravity was an accelerating force; the problem was by our every day experience we accelerate objects using contact forces, pushes or pulls; the strange thing about the gravitational force it *acts over a distance* with apparently nothing in between the gravitating masses.

Newton's Reservations about Gravity

It is important to understand that while Newton was able to formulate his law of gravity in his monumental work, he was deeply uncomfortable with the notion of 'action at a distance' which his equations implied. He never, in his words, 'assigned the cause of this power.' In all other cases, he used the phenomenon of motion to explain the origin of various forces acting on bodies, but in the case of gravity, he was unable to experimentally identify the motion that produces the force of gravity. Moreover, *he refused to even offer a hypothesis as to the cause of this force* on grounds that to do so was contrary to sound science. He lamented the fact that 'philosophers have hitherto attempted the search of nature in vain' for the source of the gravitational force, as he was convinced 'by many reasons' that there were 'causes hitherto unknown' that were fundamental to all the 'phenomena of nature.'

This fundamental phenomenon (the cause of gravity) is still under investigation and, although hypotheses abound, the definitive answer is yet to be determined.

Before we go on to theorize a possible cause and mechanism for gravity, let us discuss Einstein's view on gravity:

Albert Einstein's Gravity

Some two hundred years after Isaac Newton formulated his ideas of gravity Albert Einstein came on the scene. He was a German-born theoretical physicist who is still widely considered to have been one of the greatest physicists of all time. While best known for the theory of relativity (and specifically mass-energy equivalence, $E=mc^2$), he was awarded the 1921 Nobel Prize in Physics 'for his services to Theoretical Physics, and especially for his discovery of the law of the photoelectric effect.' In popular culture, the name 'Einstein' has become synonymous with genius. In 1999 Einstein was named Time magazine's 'Person of the Century.' Einstein's many contributions included his special theory of relativity, where he concludes *all observers will always measure the speed of light to be the same no matter what their state of uniform linear motion is* and his general theory of relativity which extended the principle of relativity to include his *geometrical theory of gravitation*.

General relativity is currently the most successful gravitational theory, being almost universally accepted and well confirmed by observations. The first success of general relativity was in explaining the anomalous perihelion precession of Mercury. Then in 1919, Sir Arthur Eddington announced that observations of stars near the eclipsed Sun confirmed general relativity's prediction that massive objects bend light. Since then, many other observations and experiments have confirmed many of the predictions of general relativity, including gravitational time dilation, the gravitational red-shift of light, signal delay, and gravitational radiation. In addition, numerous observations are interpreted as confirming the weirdest prediction of general relativity, the existence of black holes. In Einstein's view gravity was not a force as formulated by Newton, but acceleration. His general theory dictates that:

A mass in empty space-time geometrically distorts the space-time around it resulting in an acceleration of all masses in the space-time gravitational field.

In this theory, space-time is treated as 4-dimensional, the three dimensions of space and one of time, which is curved by the presence of mass, energy and momentum within it, the motion of objects being influenced solely by the geometry of space-time. I usually demonstrate the warping of space-time to my students by stretching a thin sheet of rubber over a wooden frame and placing different mass marbles in the centre. The bigger, more massive marbles cause more distortion. When a large marble is placed in the centre to represent a planet, a small marble at the edge of the sheet will promptly accelerate towards the large mass. If you're really skilled you can get the small marble to orbit the large one.

One of the defining features of general relativity is the idea that gravitational 'force' is replaced by geometry. In general relativity, phenomena that in classical mechanics are ascribed to the action of the force of gravity (such as free-fall, orbital motion, and spacecraft trajectories) are taken in general relativity to represent inertial motion in a curved space-time. So what people standing on the surface of the Earth perceive as the 'force of gravity' is a result of their undergoing a continuous physical acceleration. Their experience of weight is caused by the mechanical resistance of their accelerated motion by the surface on which they are standing. It is the curvature of space-time paths that objects in inertial motion follow i.e. 'deviate' or 'change direction' over time.

This deviation appears to us as acceleration towards massive objects, which Newton characterized as being gravity. In general relativity however, this acceleration or free fall is actually inertial motion. Objects in a gravitational field appear to fall at the same rate due to their being in inertial motion while the observer is the one being accelerated. This identification of free fall and inertia is known as the: Equivalence principle. The relationship between the presence of mass, energy, momentum and the curvature of space-time is given by the Einstein field equations.

The symbolic form:

$$G\mu\nu = 8\pi T\mu\nu$$

This elegant symbolic formulation of Einstein's general theory of relativity cannot be used for actual calculations, but it clearly shows the principle that 'matter tells space-time how to curve, and curved space tells matter how to move' (John Wheeler, Princeton University and the University of Texas at Austin) . The left side of the equation contains all the information about how space is curved, and the right side contains all the information about the location and motion of the matter. General relativity is beautiful and simple (to a physicist), but mathematically it's very complicated and subtle.

The actual shapes of space-time are described by solutions of the Einstein field equations.

Summary

1. Mass, energy, and momentum curves space-time (creates a gravitational field)

2. Objects in free-fall follow the contours (paths) of curved space-time

3. This motion of change in direction over time appears to us as acceleration towards a massive object - what we call gravity

4. All objects in curved space-time accelerate at the same rate.

Newton and Einstein Views of Gravity

Newton's laws of motion work very well for anything moving at much less than the speed of light. His law of gravity works very well for any place of weak gravity such as in the solar system. Spacecraft sent to the distant planets in the solar system arrive at their intended destinations (barring mechanical problems) within a few minutes of the expected time even after travelling for billions of kilometres over several years. Scientists use Newton's laws to guide the spacecraft to its destination, but when things are moving very fast (at a significant fraction of the speed of light) we need Einstein's formulations on light and inertial motion and that space and time can be radically changed in a very strong gravitational field. In Einstein's universe, gravity is not really a force, but acceleration. While it is true that Einstein's hypotheses are successful in explaining the effects of gravitational forces more precisely than Newton's in certain cases, *he too never assigned the cause of gravity or its mechanism* in his theories.

Einstein's equations tell us that:

Matter tells space how to curve, and space tells matter how to move.

But this new idea, completely foreign to the world of Newton, does not enable Einstein to assign the *cause of this power* of a mass to curve space-time any more than the Law of Universal Gravitation enabled Newton to assign its cause. Let us look at Newton's own words on the cause of gravity:

'I wish we could derive the rest of the phenomena of nature by the same kind of reasoning from mechanical principles; for I am induced by many reasons to suspect that they may all depend upon certain forces by which the particles of bodies, by some causes hitherto unknown, are either mutually impelled towards each other, and cohere in regular figures, or are repelled and recede from each other; which forces being unknown, philosophers have hitherto attempted the search of nature in vain. <u>If science is eventually able to discover the cause of the gravitational force</u>.'

In a way his wish came true, Einstein described the motion of objects and gravity more accurately, but did not assign a cause to gravity or its mechanism, yes mass warps space-time which we see as acceleration, but why exactly does a mass warp space time and what physical meaning does this pertain to? If a gravitational field is distorted space and distorted time, what exact physical change occurs?

We cannot simply accept that there is a distortion of the dimensions of space-time, without knowing exactly the fundamental physical structure of such dimensions and the mechanism by which a matter particle changes this physical structure. Our understanding of space-time distortion cannot be left as a pure mathematical description, it needs substance.

We need an actual physical cause of gravity other than it is just mass and it is with an understanding of *why a mass creates a gravitational field* that we ourselves may one day be able to create or modify such a field. We have already discussed the fact that in Einstein's mass-energy equivalence that mass and energy are interchangeable, that matter is condensed electromagnetic energy which can be released as electromagnetic radiation and it is here in the nature of matter itself that we may find nature's mechanism for a mass to generate a warped space-time i.e. the cause of a gravitational field.

Some Important Gravitational Effects of Space-Time Curvature

The following important effects can be used to detect the presence of a gravitational field *and* more importantly can be used to test a theoretical model for the actual mechanism of gravity:

1. Gravitational red-shifting of light

The wavelength of light increases (become 'redder') as it moves from a strong gravitational field to a weaker one. If you were to shine a blue beam of light upwards from the surface of a planet, where the gravitational field is strongest, as it moves upwards to where the gravitational field becomes increasingly weaker, the light would change colour - become redder. This is because the wavelength gets longer:

Question: why is the wavelength shorter in a gravitational field?

Answer: According to Einstein's General Relativity theory: In a gravitational field the dimensions of space-time are squashed along the vertical lines of the radial gravitational field.

This gravitational red shift *needs* a *physical explanation* and one that is based on solid classical and modern quantum physics. We cannot simply accept that the dimension of *empty space* is squashed, what exactly is squashed? How can you squeeze empty space? Part of the answer to this must be that empty space is not empty at all!

Space must have substance!

And since a ray of light leaving a gravitating mass increases in wavelength (gets redder), then:

A strong gravitational field must be 'optically' denser than a weak gravitational field.

This brings us immediately back to the notion of 'action at a distance,' such as gravity acting over empty space, a concept that Newton was uncomfortable with. The notion that geometrical space may indeed have substance Einstein himself acknowledged later on in his life. Here is a quote from his address delivered on May 5th, 1920, in the University of Leyden:

'Recapitulating, we may say that according to the general theory of relativity space is endowed with physical qualities; in this sense, therefore, there exists an ether. According to the general theory of relativity space without ether is unthinkable; for in such space there not only would be no propagation of light, but also no possibility of existence for standards of space and time (measuring-rods and clocks), nor therefore any space-time intervals in the physical sense. But this ether may not be thought of as endowed with the quality characteristic of ponderable media, as consisting of parts which may be tracked through time. The idea of motion may not be applied to it.'

2. Time Dilation

A gravitational field distorts the time dimensions of space-time. Clocks run slower in a strong gravitational field compared an observer in a weak gravitational field. The clock at the top of a mountain will run faster than a clock at sea level. Clocks in satellites in orbit around the Earth have to compensate for the time difference with clocks running on the Earth's surface. Similarly, there must be a *physical* reason for this time dilation and part of the answer lies in what we have already reasoned:

Gravitational fields slow down the speed of light

And since light governs time:

Gravitational fields must slow down time

Linking this back to what we said before: that empty space must have substance, it is this substance of empty space that is responsible for Einstein's gravitational General Theory effects on space-time.

3. The Shapiro Delay

The Shapiro time delay effect, or gravitational time delay effect, is one of the four classic solar system tests of General relativity. Radar signals passing near a massive object take slightly longer to travel to a target and longer to return (as measured by the observer) than it would if the mass of the object were not present. Signals (including light) will take longer than expected to move through a gravitational field. Dr. Shapiro was the first to make use of a previously forgotten facet of Einstein's relativity theory - that the speed of light is reduced when it passes through a gravitational field. The stronger the gravitational field the slower the speed of light.

This is similar to the reduction of the speed of light discussed earlier (refraction) when light travels through a more optically dense medium.

It seems therefore, that a gravitational field acts exactly like a transparent medium which affects the speed of light and hence time.

For example: due to the presence of the Sun's gravitational field, a radar signal travelling from the Earth to Venus and back, would have a delay of about 200 microseconds. The standard explanation for time delay in the Shapiro effect is given below:

The speed of light in meters per given interval of 'proper time' is a constant, however the travel time of *any* electromagnetic wave, or signal, moving at 299,792,458 meters per 'second' can be affected by the gravitational time dilation in regions of space-time through which it travels. This is because the coordinate time and proper time diverge as the gravitational field strength increases. The time delay (Δt) is directly proportional to the mass (M) of the object which distorts the space-time. This is what Einstein said about the Shapiro delay:

'In the second place our result shows that, according to the general theory of relativity, the law of the <u>constancy of the velocity of light in vacuum</u>, which constitutes one of the two fundamental assumptions in the special theory of relativity and to which we have already frequently referred, <u>cannot claim any unlimited validity</u>. <u>A curvature of rays of light can only take place when the velocity of propagation of light varies with position.</u> Now we might think that as a consequence of this, the special theory of relativity and with it the whole theory of relativity would be laid in the dust. But in reality this is not the case. We can only conclude that the special theory of relativity cannot claim an unlimited domain of validity ; its results hold only so long as we are able to disregard the influences of gravitational fields on the phenomena (e.g. of light).' - Albert Einstein (The General Theory of Relativity: Chapter 22 - A Few Inferences from the General Principle of Relativity.)

Following on from our previous deductions it seems that so far these physical effects of geometric warped space-time in Einstein's General Relativity can be explained by known classical effects in physics:

a. A space-time behaves like a transparent medium with optical mass-energy density and it is this energy density of the fabric of space-time which limits the speed of light to a constant (c).

b. This 'optical' medium of space is more energy dense near the surface of a gravitating mass and less dense away from the mass.

c. There must be an energy density curvature of this optical medium around a gravitating mass which decreases as an inverse square law away from the mass — thus following exactly Newton's inverse square law for gravity and Einstein's curvature of space time.

Summary

It is this changing density of the medium of space which is responsible for changes in the speed of light in a vacuum, and since the speed of light governs time, this explains the time dilation in a gravitational field and thus the Shapiro Effect. The changing density of the medium of space also explains the wavelength changes in gravitational red shift by simple refraction through optically different density media.

4. Bending of light

When light passes through a gravitational field it follows a curved path. This we can now see as the process of refraction through the substance of empty space, around the planet where gravity is strongest, the optical medium of space must be more dense causing a light ray to be bent (refracted) inwards in a curved path – just like the process of refraction in mirage formation where light is curved upwards: as the light passes through less dense (hot air) near the ground through to more dense (cold air) high above the ground it curves from less dense in towards more dense air. This refractive curvature of light is caused by its reduction in speed through the medium.

5. Gravitons

According to quantum mechanics, gravitational radiation must be composed of quanta (particles of mass-energy) called gravitons. General relativity predicts that these will be spin -2 particles. They have not yet been observed.

6. Acceleration

In a gravitational field all objects fall (accelerate) at the same rate, their acceleration is independent of mass. The time period of swing of a pendulum in a gravitational field is independent of the mass, this is a good test for gravity, and any change in the gravitational field strength can alter the time period of a pendulum. In a stronger gravitational field the pendulum will oscillate faster.

In essence we understand the properties of a gravitational field and can predict using Einstein's general theory and his equations its effects very accurately, but thus far we have no explanation as to why a mass distorts space-time or in the light of the above conclusions we can more correctly say: why a mass distorts (changes) the optical energy density of space-time. We must now take a quantum leap into the world of sub-atomic particles, for if we can understand the true nature of matter we may understand why it generates the curvature of space-time.

CHAPTER 4

The Fundamental Nature of Matter

Its threefold nature

After the Big Bang the early universe was filled with massive clouds of hot hydrogen gas and during expansion local areas of hydrogen became cooler and denser and under the attractive force of gravity, between the hydrogen atoms, the clouds collapsed to form stars. This seems like a good place to start our exploration of why a matter particle affects the optical energy density of space, thus warping the space-time matrix; even the smallest part of an element, an atom, has gravity, so it makes sense to start with the simplest, the hydrogen atom.

Atomic Structure - The Hydrogen Atom

The simplest type of atom is hydrogen which consists of a single positively charged proton in its central nucleus with a single negatively charged electron in orbit around it. This wonderful basic unit of matter is intriguing, somehow this simple hydrogen atom produces its own gravitational field, because it has mass – it curves space-time, but what is its mechanism? Let us examine the hydrogen atom in more detail:

Mass of a hydrogen atom: 1.673×10^{-27} kg

The diameter of the hydrogen atom is about: 1.06×10^{-10} m

The size of the nucleus is about: 10^{-15} m

The nucleus is 100,000 times smaller than the size of the whole atom and contains most of the mass of the atom in a very small space, it is extremely dense.

Question: what is the density of a hydrogen nucleus and how does it compare to the density of steel?

Answer: Density = Mass/Volume

Density of nucleus = mass of proton/volume of proton

Density = Mass$/(4/3\pi r^3)$

$$= 1.673 \times 10^{-27} kg((4/3\pi \times (0.0503 \times 10^{-15}m)^3)$$

$$\text{Density of hydrogen nucleus} = 9.86 \times 10^{21} kg/m^3$$

The density of steel is: $7.850 \times 10^3 \ kg/m^3$

Therefore the nucleus of a hydrogen atom is a thousand billion times denser than steel.

The electron is only 1/1800 times the mass of the proton. It was once thought that the electron orbits the nucleus much like a planet orbits the sun, as a mass particle. This was a simplistic view of the structure of the atom

Bohr Quantum Model of Atom and Electron Orbital Energies

It was known that hydrogen atoms emit specific wavelengths (colours) of light after being excited. In 1913, Neils Bohr, focusing on the particle properties of electrons, constructed a quantum model to explain this. He proposed that electrons orbited the nucleus at specific radii, also called energy levels. Electrons required specific (quantized) amounts of energy to move from one energy level to another, and emitted characteristic amounts of energy when returning to the ground-state energy levels. His model predicted that electrons were more tightly bound when they were closer to the nucleus and that atoms emitted energy when electrons dropped energy levels, moving towards the nucleus. In other words, when an electron is given extra energy (excited) it jumps from its lowest energy state (ground state or lowest orbital) to a higher energy level. This extra energy can come from the atom absorbing a photon (packet) of electromagnetic energy of a specific wavelength (frequency/energy). The electron stays in this high energy state for a fraction of a second and then drops back to a lower energy state/orbital - as it does so it emits the extra energy as a photon of light.

Each energy jump corresponds to a specific amount of energy (quantum) E. Each quantum of energy is emitted as a photon of light at a specific wavelength (colour). These specific emissions are seen as line spectra in a spectrometer.

The energy of a photon of light can be calculated using Planck's equation:

Energy Quantum of photon = Plank's constant (h) x frequency of photon (f)

$$E = h f$$

Where $h = 6.626 \times 10^{-34}$ Js

Question: how much energy is needed to jump free an electron from the hydrogen nucleus?

Answer: the measured value (ionization energy) is equal to 13.6 electron volts.

Question: how much is 13.6 electron volts equal to in joules and what wavelength of light carries this amount of energy in a photon?

Answer:

$$1eV = 1.6 \times 10\text{-}19\text{J}$$

Therefore 13.6 eV $= 13.6$ eV $\times 1.6 \times 10\text{-}19$ J/eV

$$= 2.176 \times 10\text{-}18 \text{ J}$$

This is equivalent to a photon of wavelength:

Since $E = hf$ and $c = f\lambda$

Then $E = hc/\lambda$

Therefore wavelength $\lambda = hc/E$

$$= (6.626 \times 10^{-34} \text{ Js} \times 3 \times 10^{8} \text{ m/s}) / \ 2.176 \times 10^{-18} \text{ J}$$

$$= 9.135 \times 10^{-8} \text{ m}$$

$$\lambda = 91.35 \text{ nano-metres (ultra-violet)}$$

This is the highest energy transition the hydrogen electron can make. If an electron in the ground unexcited state absorbs a photon of this wavelength (U.V.) it will free itself from the electrostatic grip of the nucleus. With a lot of assumptions and adjustments, the Bohr Model fits hydrogen pretty well, but failed for all other atoms. It was soon recognized that it was fundamentally wrong, and a new approach was needed. Today we understand that the electron 'orbits' the nucleus as an oscillating standing wave of electromagnetic mass-energy.

The Wave Mechanical Model

In the mid-1920s, Erwin Schrodinger, building on the dual nature of matter, began focusing on the standing wave-like properties of the electron. So what is a standing wave? A good example of a standing wave is a guitar string. When a string is plucked it oscillates as a standing wave - we see the string oscillating up and down with maximum movement (amplitude) in the middle (antinode) of the string and minimum movement at the ends (node). The term standing wave is a little misleading, in fact if you could video the string and play it back in slow motion you would see two waves going up and down the string and being reflected at both fixed ends of the string. When these two waves meet in the middle they add their energy to each other, this is why you see maximum amplitude in the middle. The reason it is called a standing wave, is because it is moving between fixed points - staying in one place. The natural frequency at which the string oscillates is called the fundamental frequency and is equal to half a wavelength and is the lowest energy at which it can oscillate. The fundamental frequency (f_0) gives the fundamental wavelength (λ_0). Standing waves can oscillate at higher frequencies (harmonics):

By visualizing electrons as standing waves (like guitar strings) instead of 'orbiting' particles, the distinct energy levels observed by experiments could be explained. In the circumference of a circular electron standing wave only certain numbers of whole wavelengths are allowed (energy levels). Other ones result in destructive interference and are not 'allowed.' Using this idea, Schrodinger developed a mathematical model based on wave mathematics to describe the position of electrons in an atom.

For the hydrogen atom:

Electron wave resonance n = 1

First orbital where the wavelength equals the circumference of the circular standing wave:

$$\lambda_1 = 2\pi r_1$$

Electron wave resonance n = 2

Second orbital where the wavelength equals the circumference of the circular standing wave:

$$2\lambda_2 = 2\pi r_2$$

Electron wave resonance n = 3

Third orbital where the wavelength equals the circumference of the circular standing wave:

$$3\lambda_3 = 2\pi r_3$$

For a given atom, Schrodinger's Equation has many solutions, and these different solutions (energy levels) are called orbitals. These orbitals do not describe actual orbits like Bohr's model, but, instead, solutions to a mathematical equation. The standing wave model diagram is a visualization of why, if electrons have wave-like properties like wavelength, only certain orbitals are allowed. It is not meant to say that electrons move in wavy orbits around the nucleus - they oscillate around the nucleus as a standing wave of electromagnetic energy. This idea that the electron particle (with mass) can behave like a wave is called wave-particle duality.

Wave-Particle Duality Nature
of Matter and Light

We have already said that a photon (packet) of light, an electromagnetic wave, has a specific amount (quantum) of energy - in this way it can be thought of as a particle. But particles such as electrons can also behave as waves. De Broglie showed that electrons can be diffracted - spread out like a wave when passed through small gaps. Diffraction is one of the properties of waves. He derived an equation to calculate the wavelength (De Broglie Wavelength, λ_b) of a moving particle using Planck's constant (h) and the object's momentum (mv, mass x velocity):

De Broglie Wavelength

= Planck's constant (h) /momentum (mv)

$$\lambda_b = h/mv$$

So particles (electrons) can behave as waves and waves (photons) can behave as particles. This wave-particle duality for waves and the particle-wave duality for particles reveals to us the true nature of matter and energy, matter particles behaves as waves and waves behave as matter particles, this makes sense since we have already shown the electromagnetic nature of mass-energy, remembering this energy is electromagnetic light. This reinforces the concept that matter is condensed stationary waves of electromagnetic mass-energy.

Therefore, since matter (mass) is a standing wave of electromagnetic energy

and light electromagnetic energy is a free moving wave.

It is not difficult to see how matter can be transformed

into energy (light) and vice versa.

This is Einstein's mass-energy equivalence, now $E = mc^2$ makes sense.

Students of physics sometimes ask, 'why c squared?' The simple answer is that the left hand side of the equation is units of energy, so the right hand side of the equation must reduce to units of energy and for this to happen we have to have speed squared. In terms of the fundamental physical reason, it may be something to do with the fact that the energy of a wave is proportional to the amplitude of a wave squared or the fact that the wave has both, electric field energy and magnetic field energy.

Summary for Wave-Particle Duality

Plank's equation describes light as a quantum (particle) of energy:

$$E = hf$$

The left hand side of the equation (E) a quantum (packet) of energy shows that a photon is a particle; the right hand side which includes wave frequency reminds us that it is also a wave – a photon of electromagnetic light is a wave packet.

De Broglie's equation describes particles as waves:

$$\lambda_b = h/mv$$

The momentum (mv) on right hand side of the equation reminds us that it is really a particle and the left side with wavelength shows us that a particle is also a wave.

Question: what is the De Broglie wavelength for an electron?

Answer: assuming the velocity for an electron in oscillatory orbit around the nucleus is the *speed of light* (c):

$$\lambda_b = h/mv$$

$$= 6.626 \times 10^{-34} Js \ /(9.11 \times 10^{-31} kg \times 3 \times 10^{8} m/s)$$

$$\lambda_b = 2.424 \times 10^{-12} m$$

A size, which puts it exactly in the right place in orbit in the atom: bigger than the size of the nucleus (10^{-14}m) and smaller than the diameter of the atom (10^{-10}m). The electron fits as a standing wave. If we assume the electron is in a circular orbit where the wavelength equals the circumference then the radius of the orbit:

$$Radius = circumference/2\pi$$

$$= 2.424 \times 10^{-12} m/2\pi$$

$$r = 3.859 \times 10^{-13} m$$

Therefore, a diameter equal to 7.72×10^{-13}m. This fits well for our model of the atom. We have seen already that mass and energy are interchangeable and that matter is really a standing wave of electromagnetic light energy, atoms can absorb electromagnetic wave energy as a photon which increases the orbital energy of the electron. In this case the matter particle absorbing the photon as increased its mass by:

$$Since \ \Delta E = mc^2$$

$$m = \Delta E/c^2$$

$$And \ since \ \Delta E = h\Delta f$$

Then this mass increase due to photon absorption is:

$$m = h\Delta f/c^2$$

This is energy-mass conversion, free electromagnetic wave energy travelling at the speed of light, becoming a standing wave of electromagnetic energy. Similarly the electron drops back down from a high energy state to a lower one and emits a photon packet of electromagnetic wave energy, a standing electromagnetic wave to a free one travelling at c. So by $E = mc^2$ we see that mass energy conversions are:

Light Energy to Matter: Is equivalent to: free electromagnetic wave travelling at c becoming a condensed standing electromagnetic wave.

Matter to light energy is the converse of this: this is a mass (standing electromagnetic wave energy) to free energy (electromagnetic wave packet of energy). This demonstrates that mass-energy conversions are really conversion of the mode of propagation of 'light' waves from 'free wave' to 'standing wave' and vice versa.

EXERCISE

Deriving an Equation which describes Mass-Energy Conversions

Continuing on the theme of mass-energy equivalence and wave-particle duality I recall that one afternoon while teaching three physics students about quantum mechanics and the wave-particle duality nature of light and matter, an interesting thing happened. I had just put the two equations on the board to summarize the particle nature of light and the wave nature of matter:

$$\text{Plank: } E = hf \quad (1)$$

$$\text{De Broglie: } \lambda_b = h/mv \quad (2)$$

When one of my students asked if there was an equation which linked the two? My excited reply was: 'Let's see, shall we?' We noticed that Planck's constant was in both equations, so the equations could be rearranged in terms of (h) and then equated:

$$\text{Plank: } h = E/f \ (3) \qquad \text{De Broglie: } h = \lambda_b \times m \times v \ (4)$$

$$\text{Since } h = h$$

$$E/f = \lambda_b \times m \times v \quad (5)$$

To equate particle and wave we need the velocity (v) to equal the speed of light (c) so:

$$\text{Let } v = c \quad (6)$$

Substituting v with c in equation (5) we get:

$$E/f = \lambda_b \times m \times c \quad (7)$$

Now from the wave speed equation we know that:

$$v = f \times \lambda \quad (8)$$

For light v = c the speed of light, so:

$$c = f \times \lambda \quad (9)$$

Rearranging equation (9) gives:

$$\lambda = c \, / \, f \; (10)$$

Now we substituted equation (10) back into equation (7):

$$E/f = c/f \times m \times c$$

Seeing this derived equation I was very excited about where this exercise had led to and asked my students if they realised what this equation means, one of them did.

We cancelled the f on both sides and were left with:

$$E = c \times m \times c$$

Giving:

$$\mathbf{E = m \, c^2}$$

Einstein's mass-energy equivalence! Highly excited we thought back to where we started: Einstein's equation of mass-energy equivalence ($E = mc^2$) links the two equations which define particles (mass) as waves (energy) and waves (energy) as particles (mass) - it all makes sense!

The Fundamental Equation

There is something fundamental about Einstein's mass energy equation $E = mc^2$, what exactly does it mean that the electromagnetic light energy (E) is equivalent to the mass times the speed of light squared. Why squared? Why does light in matter have mass? Light in itself has no mass only energy? A photon of light has no *rest mass* (as do particles) but it does have mass-energy! If we are to understand gravity and its mechanism we need to delve deeper into the very nature of matter, let's start with the equation by looking at the terms in $E = mc^2$.

Fundamental electromagnetic energy (E) of the wave by Planck's equation ($E = hf$) is directly proportional to the frequency (f) of the wave, if we double the frequency of the wave we double its mass-energy.

Fundamental mass (m) we say is a measure of how much matter there is – the standing electromagnetic energy waves (matter particles).

Fundamental speed of light (c) is the wave speed of both the free wave (light) and the standing wave i.e. matter, light always travels at the same speed, whether it's a free wave through the vacuum of space or oscillating at the speed of light in matter.

There are two important fundamental constants which give us the speed of light:

1. The constant of magnetic fields:

$$u_0 \text{ The permeability of free-space} =$$

$$4\pi \times 10^{-7} \text{ Wb A}^{-1} \text{ m}^{-1}$$

This constant determines the strength of a magnetic field in a particular medium, in this case the space vacuum.   For example the magnetic field strength around a current carrying conductor:

$$\text{Magnetic field strength B} = u_0(N/L)I$$

Where:

N is the number of turns of wire around the electromagnet

I is the current in amperes

L is the length of the magnetic circuit

2. The constant for electric fields:

$$\varepsilon_0 \text{ The permittivity of free-space}$$

$$= 8.8542 \times 10^{-12} \text{ Fm}^{-1}$$

This constant determines the strength of an electric field in a medium. For example the electric field strength around a point charge:

$$\text{Electric field strength per unit charge:}$$

$$E = Q/(4\pi \, \varepsilon_0 r^2)$$

Where:

Q is the charge of the particle creating the electric field in Coulombs (C)

r is the distance from the particle with charge in metres (m)

What is interesting about these constants (of the space vacuum) is that they determine the speed of light (c)!

$$c^2 = 1/u_0\varepsilon_0$$

Let's take another, very revealing, look at matter: By Einstein's equation $E = mc^2$

$$\text{Mass } (m) = E/c^2$$

Mass is energy per light speed squared!

Substituting the electric and magnetic constants we have:

$$\text{Mass} = Eu_0\varepsilon_0$$

This makes sense since matter is made out of electric and magnetic light wave energy; it shows us the electric and magnetic nature of matter. Now let us substitute the E using Planck's equation $E = hf$ and we get:

$$\textbf{Mass} = \textbf{hf}u_0\varepsilon_0$$

Considering the mass of a particle such as the electron:

$$\textbf{m}_e = \textbf{hf}u_0\varepsilon_0$$

This is a highly revealing equation about matter! It tells us that not only that matter electromagnetic in nature (the constants $u_0\varepsilon_0$), but that matter is a wave of frequency (f). But the most startling thing in this equation, and the most revealing about the nature of matter, is it that it contains Planck's constant (h).

Mass which creates a gravitational field is

electromagnetic in nature.

Summary

1. Electromagnetic waves have an oscillating electric and magnetic field which propagates (travels) at the speed of light.

2. Matter and light energy (electromagnetic waves) are interchangeable by $E = mc^2$.

3. Newton: Gravity is the force of attraction between massive particles.

4. Gravity is acceleration due to a mass curvature of space-time.

5. Mass (matter) is made up of standing waves of electromagnetic energy.

6. An electron around an atom is a standing wave of electromagnetic energy.

7. The mass particles can be defined by the fundamental mass equation:

$$m_p = hfu_0\varepsilon_0$$

CHAPTER 5
The Trouble with Matter

Having delved into the structure of matter and seen mass as a standing wave of electromagnetic light energy we now explore the quantum world of sub-atomic particles. This journey will lead us to some interesting conclusions about the very nature of matter and mass energy equivalence.

The Fundamental Structure of Matter

We understand that the atom is made up of electrons in 'orbit' as stationary electromagnetic waves around a central positively charged nucleus and that this nucleus is extremely small compared to the size of the atom and extremely dense. Here the positively charged protons balance out the negatively charged electrons making the atom (and matter) electrically neutral. The nucleus itself contains particles called nucleons. There are two types of nucleons:

Protons and Neutrons

The proton is positively charged (+1) i.e. we say it has positive unit charge, but the neutron has no net charge (0). Originally it was thought that the electron, proton and neutron were fundamental i.e. the smallest parts of matter, but in modern physics we now know that the proton and neutron are made up of other 'particles.' Collisions of these 'fundamental' particles in high energy particle accelerators have shown us that the proton and neutron have sub-structure – lumpiness! Their mass and charge are not evenly distributed into a perfect sphere.

Deep inelastic scattering experiments, where particles are smashed into each other at near light speeds provided the evidence that protons and neutrons were made up of smaller more fundamental 'particles.' Deep inelastic scattering is the process used to probe the insides of heavy particles (hadrons, such as protons and neutrons), using electrons, and closely related muons and neutrinos (leptons, light particles). The results of these experiments provided the first convincing evidence of the reality of quarks, which had previously been thought to be a purely mathematical phenomenon.

Analysis of the results of these experiments led to the following conclusions:

a. The hadrons (heavy particles) do have internal structure.

b. In baryons (protons and neutrons), there are three points of deflection i.e. baryons consist of three quarks.

c. In mesons, there are two points of deflection (i.e. mesons consist of a quark and an anti-quark.

Quarks appear to be point charges, as electrons appear to be, with fractional charges.

There are six types of quarks known: up, down, strange, charm, truth, and beauty. Each type of quark carries a specific fractional charge:

Up +2/3

Down -1/3

Charm +2/3

Strange -1/3

Top +2/3

Bottom -1/3

All of these quarks have spin ½. The proton is composed of two up quarks (u) and one down quark (d) bound together by the strong nuclear force and the neutron is composed of one up quark and two down quarks. By adding the charges of individual quarks together, we can show the charges of the nucleons they compose.

Quarks can compose particles other than protons or neutrons. However, these particles are very exotic and rare. They also usually have very short lives; they often decay after less than a second.

A proton is composed of uud (up-up-down) quarks and a neutron is made up of udd (up-down-down) quarks:

Proton charge = charges of its components added = $(+2/3) + (+2/3) + (-1/3) = +1$

Neutron charge = charges of its components added = $(+2/3) + (-1/3) + (-1/3) = 0$

Single quarks have never been observed and it is believed that quarks must exist in pairs or triplets. Quarks are thought to be fundamental particles (like electrons), which means that nothing smaller composes them. This is not proven, however and physicists believe it is possible that quarks are also composed of smaller particles. This I consider unlikely as I will explain later. Besides quarks which make up ordinary matter, there are anti-quarks (with opposite charge and 'spin') which make up anti-matter.

Antimatter

You may have watched the American TV series, *Star Trek*. The characters talk about anti-matter as the source of power for the ship. In the core of the *Warp Drive* of the Star Ship Enterprise, they mix matter and anti-matter together to provide the enormous amount of energy needed to drive the mighty warp engines. These warp engines drive the space ship at light speeds and greater by warping space and time around the ship – they ride an artificially generated gravitational wave!

So why does the mixing of matter and anti-matter provide energy? Well, it turns out that anti-matter isn't just science fiction. Anti-matter is actually science fact. But, what is it? Simply put, anti-matter is the opposite of normal matter. For every 'normal' particle of matter there is an opposite, anti-particle. There are anti-electrons, which are just like electrons but instead of a negative charge, they carry a positive charge. They are called positrons. There are anti-quarks with opposite charge to normal quarks. There are anti-protons and anti-neutrons, which are composed of anti-quarks that are just like real quarks, but have opposite properties. Anti-protons are negatively charged instead of being positively charged:

An anti-proton (-1 charge) is made up of two anti-up quarks and one anti-down quark: $(-2/3) + (-2/3) + (+1/3) = -1$

An anti-neutron (0 charge) is made up of two anti-down quarks and one anti-up quark: $(-1/3) + (-1/3) + (+2/3) = 0$

Anti-neutrons are neutral, just like normal neutrons, but have the opposite spin of a regular neutron. The anti-electrons called positrons have a positive charge. But these antimatter particles are short lived - they are soon annihilated by normal matter.

Matter Verses Antimatter

Whenever anti-matter comes in contact with matter,

both masses convert directly into electromagnetic energy, gamma rays, in a violent explosion.

Anti-particles do not last long on earth because they almost instantly annihilate themselves on meeting matter particles. However, their presence has been recorded in specially designed machines whose purpose is to create anti-matter, among other things. Antimatter-matter annihilation is the only way known for atoms to completely convert to energy – a pure mass-energy conversion. It could be a possible energy source in the future. If a 1/2 kilogram of anti-matter could be created and stored, it could be combined with a 1/2 kilogram of regular matter to release the amount of mass-energy equivalent to 1 kilogram.

Question: how much energy would be released if ½ kg of matter annihilated ½ kg of anti-matter?

Answer:

$$E = m c^2$$

$$= 1\text{kg} \times ((300,000,000\text{m/s})\text{ squared})$$

$$\text{Energy released} = 90,000,000,000,000,000 \text{ Joules of energy}$$

This is an enormous amount of energy from 1 kg of matter/anti-matter fuel! *This amount of energy could run a 1000 Watt electric motor for 5.7 million years!* However, no one knows how to make that much antimatter and if we could, how could it be stored, certainly not in a container made of normal matter. In Star Trek and the *Starship Enterprise* – the anti-matter is contained by magnetic containment so that is does not touch the matter container walls.

Matter verses anti-matter - Annihilation

When a particle and anti-particle come together they annihilate each other and release their mass-energy as two electromagnetic gamma ray photons. For example, this happens if an electron and an anti-electron meet or a proton and anti-proton. What is interesting about this mass to energy conversion is that:

From these two mass particles, which as we have seen are standing waves of electromagnetic energy, come two electromagnetic gamma light rays travelling at the speed of light.

Matter is made into light!

So annihilation is when a particle collides with an anti-particle where the electromagnetic energy is released as two gamma waves. It is these two high frequency gamma waves which is the released energy. Let us do some maths.

Question: what is the frequency and wavelength of the emitted gamma rays when an electron and anti-electron (positron) collide?

Answer: since both the electron and anti-electron have the same mass (but opposite charge) and produce 2 gamma ray photons, we calculate the mass-energy transformation for one particle:

$$\text{Mass of electron } (m_e) = 9.11 \times 10^{-31} \text{ kg}$$

By Einstein's mass-energy equivalence equation in annihilation this will produce an amount of energy (E):

$$E = mc^2$$

$$= 9.11 \times 10^{-31} \text{ kg} \times (3.0 \times 10^8 \text{ m/s})^2$$

The energy of the gamma photon

$$= 8.199 \times 10^{-14} \text{ Joules of energy}$$

(This is equivalent to 0.51MeV)

Now using Planck's equation (E = hf) we can calculate the frequency of a photon with this amount of electromagnetic energy:

$$E = hf$$

Therefore: $f = E/h$

$$= 8.199 \times 10^{-14} \text{ J} / 6.63 \times 10^{-34} \text{ Js}$$

Frequency of the gamma photon = 1.237×10^{20} Hertz (waves per second)

This frequency shows that it is in the gamma radiation part of the electromagnetic spectrum. Now let us calculate the wavelength (λ) of this photon using the wave speed formula $v = f\lambda$:

Where c is the speed of light

$$c = f\lambda$$

Therefore:

$$\lambda = c/f$$

$$= (3 \times 10^8 \text{m/s})/(1.2367 \times 1020 \text{Hz})$$

Wavelength of the gamma photon $= 2.426 \times 10^{-12}$ m!

Question: do you see something familiar about the wavelength of the gamma wave that has come from the annihilation of the electron?

Answer: It is the same wavelength as the standing wave electron around an atom, the De Broglie wavelength.

Matter from Light - Pair Production

Now there is a phenomenon in nature which is the exact opposite of mass to energy annihilation: Pair Production. Pair production can occur when two gamma wave photons of sufficient energy (frequency) meet to produce two particles, one of matter the other of anti-matter, such as an electron and positron. If the gamma rays have sufficient energy they can create heavier particles such as a proton and anti proton.

This is energy into mass conversion

Light is made into matter!

The photon need only have a total energy of twice the electron mass (i.e. 1.02 MeV) for this to occur, if it is much more energetic, heavier particles may also be produced. Gamma-ray photons with energy greater than 1.02 MeV may interact with a nucleus to form an electron-positron pair. This amount of energy is just sufficient to provide the rest masses of the electron and positron (0.51 MeV each). Excess energy will be carried away equally by these two particles which produce ionization as they travel in the material. The positron is eventually captured by an electron and annihilation of the two particles occurs. This results in the release of two photons each of 0.51 MeV known as annihilation radiation. These two photons then lose energy by Compton scattering or the photoelectric effect.

Pair Production - matter out of electromagnetic light energy!

Pair production and annihilation are the most striking examples of mass-energy equivalence (E = mc²)

Annihilation: Matter into light

Pair Production: Light into matter

In essence, matter, the standing wave, becomes light, the free wave and light, the free wave becomes matter the standing wave. This reinforces the true of the nature of matter:

Matter is made of light!

Summary

1. Electrons and quarks are fundamental particles of matter.

2. Matter and antimatter particles can annihilate each other releasing energy as gamma ray photons where:

Mass to Energy

Mass to Photon energy

$$mc^2 = hf$$

Standing electromagnetic wave to free electromagnetic wave

3. Gamma ray photons can collide and create matter and antimatter pairs:

Energy to Mass

Photon energy to Mass

$$hf = mc^2$$

Free electromagnetic wave to standing electromagnetic wave

4. Matter is a standing wave of electromagnetic energy and light is a free wave of electromagnetic energy – the two are interchangeable by $E = mc^2$.

5. Matter is made of light.

CHAPTER 6
The State of Matter
The Big Bang

There is an event in the evolution of the universe which leads us to some important evidence and conclusions about the true nature of matter and gravity, for this event we must go back in time about 14 billion years – to the Big Bang!

The Big Bang is the scientific theory that the universe emerged from a tremendously dense and hot state about 13.7 billion years ago. The theory is based on the observations which indicate the present day expansion of the universe. Extrapolated into the past, these observations show that the universe has expanded from a state in which all the matter and energy in the universe was at an immense temperature and density. There are two main sources of evidence for the Big Bang theory of the origin of the universe:

1. Red shift of light from distant galaxies

2. The cosmic background radiation of empty space

The universe exploded from a singularity releasing an enormous amount of mass-energy which created the known universe of space-time and matter

We will look at these two pieces of evidence in detail, but in order to understand let us first examine the seven possible states of matter. We are all familiar with the three states of matter: solid, liquid and gas. Water and its three states are the most familiar to us in our everyday experience:

Solid Ice - Liquid Water - Water vapour

Steam is not actually water vapour, which is an invisible gas, but condensed particles of hot liquid water. We say there is a change of state when water changes from ice into water or water into water vapour (and vice versa). To change the state of a substance from solid to liquid we have to give the substance heat energy (raise its temperature) in order to overcome the bonds of attraction between the atoms. We give the atoms/molecules more kinetic energy – they move faster and further apart.

Example: *water changes state from solid to liquid at 0^0C*

During a change of state there is no rise in actual temperature – the energy given is simply going in to overcome the bonds between the atoms. A change of state from liquid to gas also requires heat energy and a rise in kinetic energy. 'Free' air particles of air at room temperature are moving on average at 400m/s!

Example: *water changes from a liquid to a gas at 100^0C.*

The significant thing here is that as we increase the temperature of matter the particles gain more energy and at some fixed point of temperature change state.

Question: What will happen to a substance if we keep increasing the temperature?

Answer: It will go through possibly seven changes of state:

The Seven States of Matter

To keep the physics simple let's see what happens to matter as we increase its temperature:

1. Solid

Atoms packed close together and held by strong atomic bonds of shared electrons or ionic charges. State change: Solid to liquid - atoms are given sufficient vibrational kinetic energy to break free.

2. Liquid

Atomic bonds broken but atoms held loosely together by electrostatic forces (Van de Vaal's forces resulting from small negative and positive charges on the water molecule.) State change: Liquid to gas: Extra kinetic energy given to individual particles enabling their speeds to jump above escape velocity and evaporate.

3. Gas

Gas atoms and molecules move around freely at high velocities and in random directions overcoming the weak electrostatic forces between them which decrease with separation distance. State change: gas to plasma. The negative electrons in 'orbit' around the atoms are given sufficient kinetic energy to free themselves from the electrostatic attraction of the positively charged nucleus.

4. Plasma

Plasma is a high energy, high temperature soup of positive ions (nuclei of protons) and negative ions (electrons). A low temperature plasma lamp – in this case the electrons are ripped off atoms by high voltage electric field.

5. Fundamental Particle Plasma

The quarks inside the protons separate and become free in a hot soup of quarks and electrons. A proton breaks down into separate quarks. It is possible that quarks can exist as fundamental particles at such high temperatures.

6. Light - Particle Plasma

Fundamental particles and anti-particles interchange between high energy gamma ray photons and particles through annihilation and pair production.

7. Light Energy

A soup of extremely high frequency gamma rays, too hot to condense into matter – pure energy.

The higher the temperature the higher the kinetic energy (movement energy) of the particles, therefore they overcome the bonds which hold matter together, but as you can see if you keep on increasing the temperature of matter it eventually breaks down into its constituent parts and ultimately ends up as light. The point of all this is to show that the fundamental nature of matter is light. In other words matter is electromagnetic in nature. The reverse of the above process happened during the creation of the universe. At time zero the universe was at a temperature greater than 10^{32} kelvin! At this time only light energy could exist - light radiation with a frequency of 10^{43} Hertz ie extremely high frequency/energy gamma radiation. Expansion and cooling followed until matter particles condensed out of the pure energy. Eventually the temperature dropped enough for quarks to condense into ordinary 3 quark matter particles, protons and neutrons, which then condense with electrons to form the first atoms (mostly hydrogen). The evidence for the big bang comes mainly from the discovery of the red shift of light from distant galaxies and the cosmic background microwave radiation which fills all 'empty' space.'

Red Shift of Galaxies

Early in this century, astronomers noticed that distant galaxies had peculiar light spectra. Specifically, the galaxies' light spectra were shifted toward the red end of the spectrum. In 1929 astronomer Edwin Hubble compared the galaxies' spectra with their distances, calculated using different methods, and showed that the amount of red shift was proportional to distance. Hubble and others realized that the most obvious explanation for the red shift was that the galaxies were receding from Earth and each other, and the farther the galaxy, the faster the recession. This conclusion is based on a cosmological effect that is similar to the everyday Doppler shift. Doppler shift is what makes a car sound lower-pitched as it moves away from you. It turns out that a special version of this everyday effect applies to light as well - if an astronomical object is moving away from the Earth, its light will be shifted to longer (red) wavelengths. In 1948, Russian-born physicist George Gamow realised that if all the galaxies are flying apart at high speed, the entire universe must have been concentrated in a single point at some time in the past. Another scientist, Fred Hoyle, coined the name 'Big Bang' to disparage Gamow's idea, but the term became popular and has prevailed to the present day. But the Big Bang theory and the expanding universe turn out to be the best answer to Olbers' Paradox: In a stable, infinite universe, the night sky should blaze with the light of the stars that lie in all directions, even those far away. Here is why:

1. A static universe, one that is not expanding, cannot dispose of stellar energy. This kind of universe heats up over time. (We will come back to this point later, as it has some bearing on our theory of the causal agent of gravity)

2. An expanding universe continually increases its volume, which accommodates the increasing quantity of energy produced by stars; therefore the temperature of the universe does not increase.

Cosmic Background Microwave Radiation

When Gamow first formulated his theory, he predicted there would be a leftover radiation 'signature' from the Big Bang, and he realized it might be detectable. He calculated the original temperature of the explosion, took into account the temperature reduction that would be caused by the universe's subsequent expansion and arrived at a figure of about 5 Kelvin. In the 1960s Arno Penzias and Robert Wilson were working at AT&T Bell Laboratories, trying to improve microwave communications by reducing antenna noise. They found a noise in their antenna they simply couldn't remove. They considered all kinds of possibilities including bird droppings, but nothing helped. If the antenna was pointed at the sky, the noise appeared. The pointing direction and time of day didn't matter. Finally they called an astrophysicist at Princeton, who told them what the signal probably was, hung up the phone, turned to his associates and said, 'We've been scooped.' The annoying noise was, in fact, the primordial radiation left over from the Big Bang. Penzias & Wilson won the Nobel Prize for their discovery.

The cosmic microwave background radiation (CMBR also referred as relic radiation) is a form of electromagnetic radiation which fills the entire universe. It has a thermal temperature of 2.725 K black body spectrum.

Question: what is the wavelength of the cosmic background radiation?

Answer: using Wien's Law we can calculate the peak wavelength of this radiation:

$$\lambda_{peak} = 0.0028977/T$$

Where T = Absolute Temperature of the radiation

$$\lambda_{peak} = 0.00289777/2.725K = 1.0634 \times 10^{-3}\ m$$

Cosmic microwave background radiation has a peak wavelength of 0.106 cm

(microwave part of the electromagnetic spectrum)

Most cosmologists consider this radiation to be the best evidence for the hot big bang model of the universe.

There is an important statement to make here about space-time and matter, when the universe came into existence: matter, space and time came into existence together, before the Big Bang there was no such thing as time. Without matter space and time cannot exist. It is the presence of matter in the universe which dictates the size of the universe and how long it lasts. It has been calculated that if the universe was only filled with a million stars it would only last for 70 years. One lifetime for a human being, so in a sense we all owe our existence to one million stars. Another thing to note here, is that the presence of mass-energy in the universe also dictates how fast time runs.

Summary

1. Matter if heated up to extremes of temperatures breaks down into its constituent parts of fundamental particles and then ultimately light energy.

2. The universe started off with a Big Bang where matter condensed out of light energy.

3. Both of these things show us that the true nature of matter is electromagnetic, reinforcing our view that matter particles are simply standing waves of electromagnetic light radiation.

CHAPTER 7
Nature Abhors a Vacuum

In the last chapter I stated that the presence of matter in the universe allows the existence of space and time, without matter time and space cannot exist: these three are inseparable:

Matter: Space: Time

We are all familiar with the idea that space between matter, stars and planets is simply empty space – a vacuum. This concept of empty space as a child I found difficult to accept, surely I thought if space were really empty and was simply nothingness how could it exist? And if space had no 'substance' how could stars and planets take up their various positions and motions placed relative to each other? Einstein helped us in understanding space and time by describing for us its four dimensions and by saying that matter curves (warps) space and time. It must have some kind of physical substance, because its four dimensions can be physically measured, 3 of length and 1 of time. Motion through space takes time and we can measure distance by velocity and time. Space is like a matrix in four dimensions in which matter and we can exist and move. Any mass body moving through space does so under the restricting laws of physics: Newton's laws of motion. Space has unique properties: objects can move through space without any resistance to that motion provided it moves at a constant velocity (speed and direction, Newton's first law), but the moment we try and change that constant motion of speed or direction we experience a resistance to that change (Newton's third law, to every action there is an equal and opposite reaction) and because of this resistance the change requires a force proportional to the magnitude of change (Newton's second law). This resistance of an object to change its motion in empty space we call mass: this is our only measure of mass. We say that mass is a property of the object, the more substance it has in terms of the number and types of atoms, the more mass it has, but this is ignoring the medium through which it moves when we measure its mass: space itself. Mass, what we call inertial mass, must be a property that should really be defined as an object's resistance to change in its motion and this must be related to the way each atom in the object interacts with the physical matrix of space. There seems to be no interaction when in constant motion, but the moment this is changed (acceleration) there is an interaction between the atoms and the physical matrix of space which we see as resistance (mass). More on this subject later, because these ideas about inertial motion, give us a cornerstone to the true nature of gravity.

I am convinced that like matter, space must be electromagnetic in nature and that its physical nature dictates the *rate* of time itself. Let us examine this space vacuum a little closer.

The Space Vacuum

The idea that the space vacuum is not truly 'empty' but had substance is not a new idea; Newton postulated that empty space was filled with some kind of substance called the ether many years ago. These early thoughts on the space vacuum being filled with some kind of transparent undetectable ether, was based mostly on the following postulate:

Since waves need a medium to travel through and that light waves travel through the vacuum of space it was originally thought that empty space was not empty at all.

In the late 19th century the luminiferous ether (light-bearing ether) was a substance postulated to be the medium for the propagation of light. Later theories including special relativity suggested that ether did not have to exist and until recently the concept of the ether was considered an obsolete scientific theory. bvMost current physicists do not see a need to have a medium for light to propagate. The combination of relativity and quantum mechanics seems to render the concept unnecessary. However, this doesn't mean it doesn't exist (just that it doesn't have to), and there remain a number of problems in modern physics that would be simplified with such a concept. So where are we today? Does the ether exist?

Although the vast majority of modern physicists reject all ether-based theories, the intuitive appeal of a causal background for 'relativistic' effects cannot be denied.

In a paper of 1958, G. Builder concluded that 'the observable effects of absolute accelerations and of absolute velocities must be described to interaction of bodies and physical systems with some absolute inertial system. Interaction of bodies and physical systems with the universe cannot be described in terms of Mach's hypothesis, since this is untenable.

There is therefore no alternative to the ether hypothesis.

So if there is evidence and a need for the space vacuum to be filled with an ephemeral substance called the ether, what exactly is it like? Let us examine some modern physics based on quantum mechanics concerning the problem of 'empty' space:

A True Vacuum

A vacuum is a volume of space that is essentially empty of *all* matter, so that gaseous pressure is much less than standard atmospheric pressure. The root of the word *vacuum* is the Latin adjective *vacuus* which means 'empty,' but space can never be perfectly empty. A perfect vacuum with a gaseous pressure of absolute zero is a philosophical concept that is never observed in practice, not least because quantum theory predicts that no volume of space is perfectly empty in this way. Physicists often use the term 'vacuum' slightly differently: the *space vacuum*. Much of outer space has the density and pressure of an almost perfect vacuum. It has effectively no friction, which allows stars, planets and moons to move freely along ideal gravitational trajectories. But no vacuum is perfect, not even in interstellar space, where there are a few hydrogen atoms per cubic centimetre.

The Space Vacuum

Space is not truly empty; it is filled with all sorts of particles and radiation:

1. Cosmic radiation from deep space: gamma rays, x-rays, etc.

2. Cosmic particles from living and exploding stars: neutrinos and other exotic particles.

3. Solar wind of charged particles from our own sun.

The entire observable universe is in fact filled with large numbers of photons, the so-called cosmic background radiation. The current temperature of this radiation as we have seen is about 3 K, or -270 degrees Celsius. The result of all this is mass-energy in the 'emptiness' of space means that there is a space vacuum pressure (P_v) generated by the photons and numerous particles zooming around the universe. But that's not the full story.

Virtual Particles and Zero Point Energy

If using a super vacuum cleaner we were to suck out all the particles and thermal heat radiation from a box insulated from all this cosmic radiation until its internal temperature reached absolute zero (0K), the 'vacuum' inside would still not be empty. The reason that a perfect vacuum is impossible is the Heisenberg's uncertainty principle which states that no particle can ever have an exact position. Each atom exists as a probability function of space, which has a certain non-zero value everywhere in a given volume. Even the space between molecules is not a perfect vacuum. More fundamentally, quantum mechanics predicts that vacuum energy can never be exactly zero. The lowest possible energy state is called the zero-point energy (ZPE) and consists of a seething mass of virtual particles that have a brief existence. This is called vacuum fluctuation. Vacuum fluctuations may also be related to the so-called cosmological constant in the theory of gravitation, if indeed this entity were to be observed in nature on a macroscopic scale. The evidence for vacuum fluctuations is the Casimir effect and the Lamb shift. Before we talk about these two effects as evidence for vacuum energy let's see exactly what we mean by virtual particles:

The Virtual Particle World of Empty Space

General relativity states that there is a zero gravitational field in empty space, but:

Quantum mechanics say it <u>averages zero</u>,
and fluctuates more and more wildly on a smaller
and smaller scale.

According to quantum mechanics, if you could magnify empty space enough, you would find that it is not flat at all but tangled, distorted, bubbly, and tumultuous. This frenzy is called quantum foam.

On a magnified scale empty space is not flat at all but averages zero energy – the quantum foam

Empty space is not flat but on the quantum level is a sea of distorted space-time. Virtual particles can come into existence, borrow energy from the space vacuum, provided they pay back this energy in a very short time, i.e. particles can come in and out of existence from empty space. Heisenberg's uncertainty principle tells us how long these particles can exist on borrowed energy: the bigger the mass-energy (Δm) of the virtual particle the shorter the time (Δt) it can exist. These particles on the quantum level are very real while they exist, but because they soon disappear back to the average zero energy of the space vacuum they are regarded as virtual:

Think of water particles jumping out of a choppy sea, they soon fall back to the water.

Heisenberg's uncertainty principle relates mass and time with the following equation:

$$\Delta m \times \Delta t = h/2\pi c^2$$

Where h is Planck's constant and c is the speed of light

The result of Heisenberg's uncertainty principle is astonishing: it means that over a short interval of time we cannot be sure how much virtual matter, mass-energy, there is in any location, even in 'empty space!' No particle can appear spontaneously by itself, for every particle of matter there must be an anti-particle created, in other words there is an equal amount of matter and anti-matter coming in and out of existence in empty space.

Question: how long can a pair of virtual electrons (electron and positron) exist for?

Answer: Using the above equation and noting the combined mass of the pair is 2 times the mass of an electron:

$$\Delta t = h/(2\pi c^2 \Delta m)$$

$$= 6.626 \times 10^{-34} \, Js/(2\pi \times (3 \times 10^8 m/s)^2 \times 2 \times 9.11 \times 10^{-31} kg)$$

$$\Delta t = 6.43 \times 10^{-22} \text{ seconds}$$

An extremely short time! A proton/antiproton pair, which is 1830 times as massive as an electron, can only exist for 1830 times less time than an electron-positron pair. The evidence for these vacuum fluctuations, virtual pairs, comes from the Casimir effect and the Lamb shift.

The Casimir Effect

The Casimir effect is the attractive force between two flat mirrors in a vacuum of 'empty' space.

If you take two flat mirrors and arrange them so that they are facing each other in empty space, you might expect nothing to happen, but in fact both mirrors are mutually attracted to each other by the simple presence of the space vacuum. This is further proof that space is not empty.

Explanation: Remember we used to think that a vacuum was what remained if you emptied a container of all its particles and lowered the temperature down to absolute zero, but since the arrival of quantum mechanics we now know that a vacuum has quantum fluctuations of energy – energy borrowed from the vacuum. All fields - in particular electromagnetic fields - have fluctuations. In other words at any given moment their actual value varies around a constant, mean value. Even a perfect vacuum at absolute zero has fluctuating fields known as 'vacuum fluctuations', the average energy of which corresponds to half the energy of a photon:

$$\text{Energy of photon}/2 = hf/2$$

Vacuum fluctuations cause an electron to shift its energy state ('orbit' or higher frequency standing wave). So what exactly causes the force between the two plates, well outside of the mirrors in free space radiation can oscillate at any frequency – a spectrum of colours, but inside between the two mirrors only certain frequencies/wavelength are allowed. This is because the radiation reflects back and forth between the two mirrors setting up a standing wave which can only exist in series of half wavelengths. The amplitude of the standing wave is twice the amplitude of the waves bouncing back and forth, this is because when they meet each other their energies add.

The field resonates in the cavity between the plates. Outside the mirrors in the vacuum the free electromagnetic radiation, which can oscillate at all frequencies, generates a field radiation pressure on the plates. This radiation pressure increases with the energy - and hence the frequency - of the electromagnetic field. If the mirrors are separated by a whole number of half wave lengths, and are therefore at a cavity-resonance frequency, the energy of the waves add to each other and the radiation pressure inside the cavity is stronger than outside and the mirrors are pushed apart. When the distance between the mirrors is not exactly equal to half wave lengths the waves cancel each other out, a crest of the wave meets the trough of another wave – in this situation it is out of resonance. When out of resonance the radiation pressure inside the cavity is smaller than outside and the mirrors are drawn towards each other. As the gap between the plates is narrowed, fewer waves can contribute to the vacuum energy and so the energy density between the plates falls below the energy density of the surrounding space. The result is a tiny force trying to pull the plates together – a force that has been measured and thus provides proof of the existence of the quantum vacuum filled with virtual particles and photons

The radiation pressure is greater outside the plates than in the cavity – pushing the plates together. The same thing happens with the virtual particles of free space, here only certain particles with a specific energy/frequency can exist as standing waves inside the cavity.

The force, F, is proportional to the cross-sectional area, A, of the mirrors and increases 16-fold every time the distance, d, between the mirrors is halved:

$$F = - A/d^4$$

Apart from these geometrical quantities the force depends only on fundamental values – Planck's constant and the speed of light:

$$F = \underline{h}c\pi^2/240d^4$$

Where $\underline{h}$ here is the reduced Planck's constant:

$$= h/2\pi$$

Since the strength of the force falls off rapidly with distance it is only measurable when the distance between the objects is extremely small.

The Casimir effect confirms the presence of a quantum vacuum where 'empty space' is filled with the energy of virtual photons and particles.

Lamb Shift

Lamb shift is the slight shift in the energy and therefore shift in wavelength and frequency, of an electron bound to a nucleus due to energy fluctuations in the space vacuum. This shift in the electron's energy level is evidence for the presence of virtual pairs of particles, such as the electron and positron or virtual photons. The electron in its 'orbit' around the nucleus wobbles because of these energy fluctuations – this effect which is seen and measured is called Zitterbewegung motion.

Lamb Shift Mechanism: between the electron and the space vacuum pairs of virtual particles and photons there is a continuous exchange of energy, therefore the electron energy levels fluctuate above a theoretical fixed level, this is seen as a broadening of the lines (wavelengths) in emission spectra, as opposed to of the emission lines being fixed at a particular quantum energy level in a vacuum devoid of energy.

Researchers at the Max Planck Institute for Quantum Optics in Garching, Germany have determined experimentally that the Lamb effect should cause the energy of an electron in the lowest energy state, the 1S state, to be shifted upward by 8172.86 MHz for the hydrogen atom and 8184.00 MHz in the deuterium atom. This energy-frequency shift is equivalent to a wavelength shift of:

$$\lambda = c/f$$

$$= (3 \times 108m/s)/(8172.89 \times 106 \text{ Hz})$$

$$= 0.03671m$$

$$= 3.671cm$$

And is equivalent to a black body radiation temperature of:

$$T = 0.0028977/\lambda_{peak}$$

$$= 0.0028977/0.03671m$$

$$= 0.0789K$$

Thinking of this in terms of energy, the energy of the particle absorbed to shift the energy of the electron by this amount is:

$$E = hf$$

$$= 6.626 \times 10^{-34}Js \times 8.17289 \times 10^{9}Hz$$

$$= 5.4154 \times 10^{-24}J$$

Thinking of this in terms of mass: the mass of the energy particle absorbed to shift the energy of the electron by this amount is:

$$m = E/c^2$$
$$= 5.4154 \times 10^{-24}J/(3 \times 10^8 m/s)2$$
$$= 6.0171 \times 10^{-41} kg$$

In terms of the uncertainty principle if this energy shift is caused by a virtual particle pair of two times this mass then its maximum time of virtual existence - before absorption by the electron is:

$$\Delta t = h/2\pi c^2 \Delta m$$
$$= (6.626 \times 10^{-34}Js)/(2\pi \times (3 \times 10^8 m/s)^2 \times (2 \times 6.0171 \times 10^{-41} kg))$$
$$\Delta t = 9.737 \times 10^{-12} seconds$$

This is a much longer than the time for an electron positron pair to exist: $\Delta t = 6.43 \times 10^{-22}$s. This is because it is less massive (electron 9.11 x 10^{-31}kg). These are important calculations in our understanding of a possible mechanism for gravity and we shall come back to them in a later chapter.

This modern quantum mechanical view that the space vacuum is not empty, evidenced today by the Casimir Effect, the Lamb Shift and Quantum Field Fluctuations of virtual particles and photons, re-enforces the idea of a substance, which we may call the ether which pervades all space and matter. In simple terms space is not empty, but filled with an immeasurable amount of energy. Some physicists have calculated that the energy density ($E\rho_v$) of the space vacuum is, on the conservative side absolutely enormous, *to being infinitely large*! In future in this book we will refer to the energy (ether) which fills the space vacuum as the: space vacuum energy (E_v)

Summary

1. The existence of the Space Vacuum Energy is evidenced theoretically by quantum mechanical fluctuations which allow virtual particles borrowing energy from Zero Point Energy (ZPE) of the space vacuum to exist for very short times without breaking any laws of physics – this mass-energy is immeasurably high.

2. The existence of the Space Vacuum Energy is evidenced experimentally by the Casimir Effect force between two reflecting

plates in an 'energy less' vacuum and the Zitterbewegung wobble (Lamb Shift) of electrons in orbit around atoms.

3. Final conclusion:

Space is not empty but filled with an enormous, probably infinite, amount of energy.

CHAPTER 8

All in a Spin

The Fundamental Mass Particle – The Electron

This quantum journey into understanding space-time, light, mass-energy and gravity is about to take another leap – back into the structure of matter. If we are to find a mechanism for gravity we need to delve deeper into the very structure of matter, after all it is matter which causes gravity, it is this substance which produces a gravitational field around an object and the observed gravitational acceleration of other masses towards it. In order to simplify this investigation I decided to look at one of the most fundamental mass particles in nature: the electron. A simple electron has mass (9.11×10^{-31}kg) and therefore gravity. If an electron can generate a gravitational field, then it must have a mechanism for doing that. The answer that it simply has mass and that mass curves space-time is not a good enough answer, this only explains what it does (Einstein's general theory of relativity) but not how it does it!

The gravitational field acceleration 'g' at a point (P_r) around an electron is very small, because its mass is very small, but for future reference let us calculate it. We will calculate 'g' at a point which is very close to the electron: a distance (r) which is its own wavelength (λ) away from its centre, so $r = \lambda$. Using our previously calculated value for the electron's wavelength ($\lambda = 2.426 \times 10^{-12}$ m) and Newton's gravity equation, then 'g' at Pr equals:

$$g = Gm_e/r^2$$
$$= 6.673 \times 10^{-11} \text{ Nm}^2\text{kg}^{-2} \times 9.11$$
$$\times 10^{-31}\text{kg}/(2.426 \times 10^{-12}\text{m})^2$$
$$g = 1.033 \times 10^{-17} \text{ m/s}^2.$$

Okay, so far so good, so what of its actual structure? If there is a mechanism then it must be related to its structure? What does an electron actually look like? Let's start with what we know: We know it has mass, charge and spin and that it is a standing wave of electromagnetic energy of calculated frequency (f).

Mass: 9.11×10^{-31} kg

Wavelength: 2.426×10^{-12} m

Frequency of electromagnetic oscillation:

1.237×10^{20} Hertz

Mass-energy equivalence: 8.199×10^{-14} J

Electric charge: 1.602×10^{-19} C

Spin: 1/2

All of the above dimensions are fairly understandable, except for spin. So what is electron spin and why does it have a value of one half? In the standard model' of electron spin, the direction of rotation of the electric charge affects the direction of the magnetic field dipole (N and S). An electron has a magnetic dipole N and S since it has charge – it is electro-magnetic in nature. The concept of electron spin has always bothered me, as most textbooks do not even attempt to explain it and say that we should be very careful with the classical interpretation of what electron spin really is. We are all familiar with the classical view of spin: the world spins on its axis, a bicycle wheel spins around and so does a spinning top, but we are warned away from this view of the electron spin. We are told that the property of electron spin must be considered to be a quantum concept without detailed classical analogy and that electron spin (s) = 1/2 is an intrinsic property of electrons. The concept of electron spin was discovered by S.A. Goudsmit and George Uhlenbeck in 1925. As far as we can understand, electron spin is the electron's intrinsic angular momentum.

In physics, the angular momentum of an object rotating about some reference point is the measure of the extent to which the object will continue to rotate about that point unless acted upon by an external torque (force). Angular momentum is an important concept in both physics and engineering, with numerous applications. For example, the kinetic energy stored in a massive rotating object such as a flywheel is proportional to the square of the angular momentum.

A gyroscope remains upright while spinning due to its angular momentum

Ordinary momentum is a measure of an object's tendency to move at constant speed along a straight path. Momentum depends on speed and mass:

$$\text{Linear Momentum (Kg.m/s)}$$

$$= \text{mass (kg) x velocity (m/s)}$$

$$L = mv$$

A train moving at 20 mph has more momentum than a bicyclist moving at the same speed. A car colliding at 5 mph does not cause as much damage as the same car colliding at 60 mph.

Angular momentum measures an object's tendency to continue to spin:

Angular momentum = mass (kg) $\times$ velocity (m/s) $\times$ distance mass is from centre of spin (m)

$$L_a = mvr$$

The units for angular momentum (of spin) are:

$$\text{Kg x m/s x m}$$

$$L_a = Kgm^2s^{-1}$$

These are extremely interesting units of spin for the electron in terms of a possible gravity mechanism.

Before we investigate the significance of the units of angular momentum in relation to the cause of gravity let's return to the electron's spin and structure.

The Electron Structure

Physics has for some time been struggling to find an accepted model of electron structure. The classical view that the electron is simply orbiting the nucleus as a point charge caused considerable problems for theoretical physicists, one being:

'If the electron is in orbital acceleration (constant circular motion where its velocity direction is continually changing) around the nucleus as a charged particle then why does it not radiate away all its energy (like other accelerating charges) and crash into the nucleus?'

It seems by the classical model that the electron does not fit standard physics, this problem was 'overcome' by quantum mechanics where we see the electron as a standing wave of electromagnetic energy – not an orbiting charge. There are many who disagree with this view saying that it does not account for all the properties of an electron.

A Fundamental Problem

Some physicists argue that the problem of the accelerating electron around the nucleus of the atom has not gone away and that quantum physicists are hiding behind this quantum view of the electron as a standing wave or refusing even discuss the problem, but if other physicists are correct *that the electron does radiate energy* then this creates another major problem and one which is much more important and consequential to the existence of the universe itself.

Consider this very important point: If the electron is actually radiating away its energy then it will be unable to sustain its spin – the consequences of this are enormous - matter in the universe cannot exist longer than a mere fraction of a second, but it does, infinitely! And here is another gargantuan problem as a consequence: if the electron really is radiating away its energy where does it get it from?

Let us look at a few models for the structure of a free independent electron away from the nucleus:

The Electron as a Spinning Ring

The electron is still regarded by many physicists as a point like particle, with no internal structure and no physical size, but how many have asked can a point particle, without any physical size, 'spin' and have intrinsic angular momentum? The 'spinning' of the point particle is meaningless. What matters is where the intrinsic angular momentum originates from inside the electron.   Some physicist reject the point charge model of the electron and have it has a spinning ring of charge:

This model proposed by Bergman and Wesley can be found in full on the internet, here we will investigate the main themes of this model.

In this model the uniform surface charge density over the ring equals the total charge e (-1.6×10^{-19}C) of the electron, where:

$$R = radius$$

$$r = half\ thickness$$

$$w = angular\ velocity\ of\ its\ spin$$

The mass of the electron is obtained from the classical electromagnetic energy of the charged spinning ring and mass-energy equivalence. The total electromagnetic energy of the spinning ring E_{em} is given by the electrostatic energy (E_e):

$$E_e = e \text{ squared}/2C$$

Where C is the capacitance of the ring

This model is beginning to make sense as the finite structure of the electron, but for a model of the electron to be correct it must account for all its physical properties including its ability to create a gravitational field – remember it is not enough any more to say because it has mass-energy (in the above case electromagnetic) and therefore it has gravity – we need to understand why and how it does this and this must be related to its structure mechanism. If we can determine how a fundamental particle like an electron generates gravity then we will be able to understand how *all* particles do it.

Properties of the Spinning Ring Electron

1. It has a tangential speed c (speed of light).

2. The ring is not a material ring, so no matter or mass is travelling with the speed c.

3. The model consists of electric and magnetic fields only and all electromagnetic fields necessarily propagate with the velocity c in a vacuum.

The speed of the ring c (speed of light) is actually the velocity of the electromagnetic field

4. The radius R of the ring equals the rationalized Compton wavelength, 3.86×10^{-13} meters (1000 times smaller than a hydrogen atom).

5. The model is completely stable under electromagnetic forces alone.

6. The frequency of rotation equals the Compton frequency.

7. The angular momentum (L_a) of a free electron depends upon its magneto-static energy (E_B) only and not upon its electrostatic energy (E_E), since only a moving charge will produce momentum, therefore:

The magnetic angular momentum alone of the spinning electromagnetic field gives the electron its inertial mass (m).

Summary of the Spinning Ring model of the Electron

The free electron spins at the speed of light as an electromagnetic ring where its magnetic angular momentum gives the electron what we see as its inertial mass.

This model begins to make sense: the electromagnetic field energy of the electron, which has no rest mass, can spin at the speed of light and what we see and measure as the inertial mass (m = 9.11 x 10^{-31}kg) of the electron comes from its magnetic angular momentum (L_B). Let us look closely at the electron's angular momentum as this may be crucial to a theory of gravity:

Angular Momentum (L_s) of Electron Spin

It is important to understand that an electron possesses an intrinsic (its own) angular momentum (the momentum of a spinning body) in addition to its orbital angular momentum about a nucleus as though it were a spinning rigid body.

The observable magnitude of this spin angular momentum is:

$$L_s = h/2$$

Where h is Planck's constant.

This angular momentum of spin (Ls): is the intrinsic spin angular momentum of the free electron and should not be confused with the spin of an electron bound in an atom and coupled to other atomic particles. We have already said that the angular momentum (and hence its inertial mass) of a free electron depends upon its magnetic energy only and not upon its electric energy, since only moving charge will produce momentum – a charged non-spinning ring would have electrostatic energy, but no angular momentum.

The spinning ring has the tangential velocity c

So that the spin angular momentum is given by:

$$L_s = m_m cR$$

Where m_m is the mass equivalent by E = mc^2 of the magnetic energy and equals:

$$m_m = m_e/2$$

In other words the magnetic mass-energy is half the inertial mass of the electron. The fact that only magnetic mass (m_m) is involved and the magnetic mass is one half the total mass of the electron means that the angular momentum of the electron L_s, is only one-half the amount that would be associated with an ordinary spinning macroscopic body.

So in answer to an earlier question what is electron spin? It is simply:

The electron spin is the electron's
electromagnetic field angular momentum

Where the observable magnitude of this spin angular momentum is:

$$L_s = h/2$$

Where h is Planck's constant.

Let us do some calculations for the free electron to reinforce these concepts about the electron's structure:

Mass of electron: me = 9.1094×10^{-31} kg

Spin speed of electron: c = 2.998×10^8 m/s

Planck's constant: h = 6.6261×10^{-34} Js

The radius of ring equals the rationalized Compton wavelength:

$$R = h/m_e c$$
$$= 6.6261 \times 10^{-34} \text{ Js} / (9.1094 \times 10^{-31} \text{ kg} \times 2.998 \times 10^8 \text{ m/s})$$
$$R = 2.4263 \times 10^{-12} \text{m}$$

The spin angular momentum equals:

$$L_s = m_m cR$$
$$L_s = (m_e/2)cR$$
$$= (9.1094 \times 10^{-31} \text{kg}/2) \times (2.998 \times 10^8 \text{ m/s} \times 2.4263 \times 10^{-12} \text{m})$$
$$L_s = 3.313 \times 10^{-34} \text{ kgm}^2\text{s}^{-1}$$

What is extremely significant about this value and the formula ($L_s = h/2$) for the angular momentum of the free spinning electron is that it *is exactly half the value of Planck's constant.*

Also, as stated above, since:

$$L_s = h/2$$

And since $L_s = (m_e/2)cR$ equating the two equations by Ls gives:

$$Ls = (m_e/2)cR = h/2$$

Then the inertial mass of the electron equal to:

The inertial mass of the spinning electron:

$$m_e = h/cR$$

This equation is most significant in that it demonstrates that the inertial mass of a particle is directly proportional to Planck's constant.

Mass of a particle α h

The mass is also inversely proportional to the speed of light (c) and its spinning ring radius (R).

Summary of Electron Structure

Here are some summary findings based on a spinning ring model for the electron:

1. Electrons have an electromagnetic origin.

2. Electron spin is the electromagnetic field's angular momentum.

3. An electron's self-energy is a result of the electromagnetic field energy.

4. An electron's magnetic field is similar to that of a magnetic dipole field, where the North Pole is a single unit with a negative magnetic charge and the South Pole is a single unit with a positive magnetic charge. Multiples of the electric charge unit 'e' and the magnetic unit 'g' equals Planck's constant 'h.'

5. Protons and neutrons have an electromagnetic origin.

6. All materials have an electromagnetic origin

Relevant Points about the Electron Structure

1. Angular momentum of spin is what we see as mass – the electromagnetic energy of the spinning ring has no rest mass in itself only electromagnetic mass-energy (hf).

2. The inertial mass of a particle depends on the constant c the speed of light – as in Einstein's $E = mc^2$.

If the speed of light increases the mass of the particle decreases – in a gravitational field the speed of light decreases and therefore mass increases!

3. The radius of the spinning ring decreases as the mass increases – thus more massive particles like the proton have a smaller spin radius.

Remembering the significant formula we derived before:

$$m = hf\mu_0\varepsilon_0$$

And that this formula supports the view that matter is electromagnetic in nature and that a particle is:

A standing wave of electromagnetic energy (light)

4. The particle mass has frequency (f) – a wave property which determines its electromagnetic energy (E = hf).

5. The particle mass has an electromagnetic wave nature by the two constants μ_0 (oscillating magnetic field) and ε_0 (oscillating electric field).

6. The fact that Planck's constant is present in the derived mass equation is intriguing because the units of this fundamental constant are joule-seconds (Js), <u>which are the units of angular momentum!</u>

Proof:

$$Js = \text{Energy} \times \text{time}$$

Since energy equals force x distance:

$$Js = \text{Force} \times \text{distance} \times \text{time}$$

And since Force equals mass x acceleration

$$Js = \text{mass} \times \text{acceleration} \times \text{distance} \times \text{time}$$

Converting to base units:

$$Js = Kg \times m/s^2 \times m \times s$$

Which gives: $Js = kgm^2s^{-1}$

So the units of Plank's constant in the mass formula ($m = hf\mu_0\varepsilon_0$) are the same as the units in the angular momentum spin formula ($L_a = mvr$): $L_a = Kgm^2s^{-1}$

This link that mass is expressed by units of angular momentum (spin) comes from:

$$m_e = h/cR$$

(Spinning ring model)

$$m = hf\mu_0\varepsilon_0$$

(Mass-energy equation $E = mc2$)

7. Planck's constant is present in both equations giving insight into the fundamental nature of matter particles:

<u>A matter particle is a spinning electromagnetic wave which we see as a standing wave travelling at the speed of light.</u>

This wave-particle, remember, has wavelength, by the De Broglie particle-wave duality equation:

$$\lambda_b = h/mc \text{ (wavelength of a mass particle)}$$

Where Plank's constant is again present confirming the angular momentum of spin for matter. A point also worth remembering from the electron spinning ring equation: $L_s = (m_e/2)cR$ is that the particle's inertial mass come from the magnetic moment of its spin:

$$m_e = 2L_s/cR$$

It makes sense to see this spinning ring of electromagnetic energy as our standing electromagnetic wave spinning around at the speed of light. It also makes sense of particle annihilation, when the mass of the particles is directly converted to pure energy as electromagnetic gamma rays. Let us revisit mass-energy conversions with our model of matter as a spinning standing wave:

Pair Production: two free electromagnetic waves with no rest mass, only electromagnetic energy, travelling at the speed of light in a straight line meet and interact to become two standing particle waves (matter and anti-matter particles) spinning at the speed of light where their angular momentum of spin gives them rest mass.

Annihilation: When a matter particle and anti-matter particle (with opposite spin) meet and interact to uncoil each others electromagnetic wave and to release them as free waves at the speed of light.

What is interesting to note is that light whether in a mass particle or as a linear free wave is always travelling at the speed of light. It is Einstein's constant remember: light can do nothing else but travel at the speed of light, whether as matter or as light energy. Einstein's mass-energy equivalence also makes more sense when we see matter

(m) as standing waves (electric and magnetic) and energy (E) as free waves (electric and magnetic):

$$E = mc^2$$

No wonder the constant the speed of light squared (c^2) is in this famous equation.

CHAPTER 9
A Matter of Flux

A pulsar, is a rapidly rotating neutron star and is one of the most powerful energy systems of the universe

The fact that matter is electromagnetic in nature, like light energy, and that the space vacuum is not empty, but filled with an unquantifiable amount of mass-energy is fundamental in our understanding a mechanism for gravity. One of the important conclusions in the last chapter was that the inertial mass of a particle is a result of its electromagnetic spin angular momentum and that this is directly related to Planck's constant h which has units of angular momentum. The equation which describes the electromagnetic spin nature of matter:

$$m = hf\mu_0\varepsilon_0$$

Matter particles have frequency (f) which describes their standing wave nature and the electric and magnetic constants demonstrates the electro-magnetic property of the wave, whereas Planck's constant with units of angular momentum describes the spin nature which is responsible for what we see as mass.

Question: what is the structure of this spinning standing wave model and for the electron, how does this structure account for a single negative charge and a magnetic dipole?

Answer: this will take some time:

The universe is filled with opposites, for example:

Magnetic south and north poles

Positive and negative electric charges

And as we know opposites can cancel each other out: for instance the atom is electrically neutral because it has equal number of positive charges (protons) and negative charges (electrons). Another interesting pair of opposites is:

Gravitational Potential Energy (GPE) – energy stored in a mass as it is raised against gravity

Kinetic Energy (KE) – the movement energy gained when a mass falls under gravity.

The interchange of energy between the two is demonstrated when you throw a ball into the air and watch it fall back down. The rising and falling ball has no energy of its own! It was borrowed from the person throwing it and returns it back (assuming you catch it). While the ball is in motion at all points on the way up and down the total energy of the system is a constant:

On the way up:

GPE gained = KE lost

On the way down:

KE gained = GPE lost

This is a good example of borrowed energy. The universe is also a very good example of borrowed energy, its expansion can be likened to a ball thrown into the air, the energy of the Big Bang gave matter the kinetic energy to 'rise' against universal gravitational and so over time the initial kinetic energy of the Big Bang is slowly being converted into gravitational potential energy and just like the ball thrown into the air, it will return providing there is enough matter and therefore enough gravity to pull everything back together. In this sense the universe is living on borrowed energy – something came out of nothing, just like virtual particles in the space vacuum.

We see this interchange of gravitational potential energy into kinetic energy when enormous cold gas clouds of hydrogen in space (nebula) collapse under gravity and form hot stars:

Cold gas with GPE to collapsing gas with KE (moving gas particles with kinetic heat energy)

Matter as Standing Waves of
Electromagnetic Energy

Let us return to the fundamental structure of matter in the universe and the fine tune our spinning ring model into a real standing wave. If an electron is a fundamental matter particle then its structure is the key to understanding a mechanism for gravity.

Electron Matter Particle

A standing wave of electromagnetic wave energy spinning with angular momentum at the speed of light (c).

How can this model explain the fact that an electron carries a single negative charge, for to make a negative we need a positive? So where is the positive electric field for the electron? The positive field must be still present but in such a way it does not cancel the out the negative field.

The answer to this question must be that the positive field is somehow hidden inside the electron and cancelled out, this is a possibility because electric fields (and magnetic) can cancel each other if they are in opposite direction and with equal magnitude.

We know that the electric and magnetic fields in a light wave have direction and are at right angles to each other and since the wave is oscillating at (f) the direction of these fields is changing every half wavelength as stated before:

Oscillating electric and magnetic fields in light: the fields are continually changing direction.

Below is a simple model of an electron standing wave where the E-field (electric) of the wave oscillates in and out:

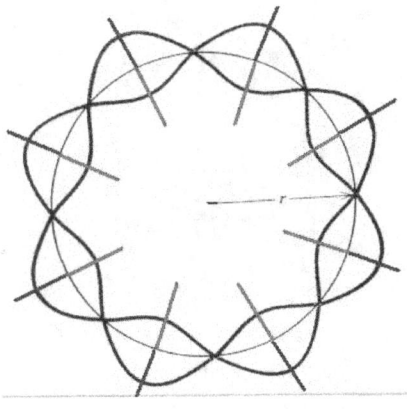

Electric 'charge' field lines in a simple circular standing wave:

blue = field out and red = field in.

In this simple model the direction of the electric field is always the same outside the wave particle which we seen as a single charge (-) where the opposite (+) field direction is hidden and facing inside. So we 'see' an electron as a single charge. The field direction is always pointing away from the particle.

This model is very simplistic it does not take into account the magnetic dipole of the electron. The configuration of the standing wave must be in such a way that the oscillating magnetic field lines of the wave, which are at right angles to the electric field (in and out of the paper) always points in the same direction. This is needed to create a constant magnetic dipole of N and S. In the above example the magnetic field would be seen to be oscillating in and out of the plane of the paper. The electron wave as it spins needs to turn inwards every half wavelength to keep the magnetic field in the same direction. This can be done if we simply twist the electromagnetic wave of the electron as it spins. In a more refined model below the electron electromagnetic wave is a doughnut shaped ring where the wave is travelling at c anticlockwise and making one complete twist of the cylinder per revolution – which equals one whole wavelength. In this case the circumference equals the wavelength and the field is always out.

Doughnut ring model of an electron: a single wavelength electromagnetic standing wave where the wave travelling at c makes one complete revolution around the ring while simultaneously rotating as it spins.

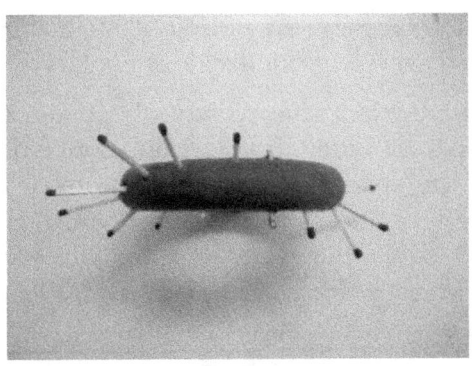

Electron model viewed edge on showing that because the wave is revolving around the cylinder the electric field is always out

In this model the magnetic field (B) is always in the same direction – UP! This configuration gives the electron it's a magnetic dipole of N and S. The North Pole is above the ring and the South Pole is below it. Whatever the convoluted shape of the electron standing wave, it is the way it folds and spins which gives rise to its properties.

This model for the electron is similar to smoke rings which are simple revolving toroids. The simple smoke ring has some remarkable properties: smoke rings can move through air at a constant velocity with very little drag – it is only when the smoke ring breaks down does it suddenly slow down and stop. A simple smoke ring can be blown across the full length of a room before in breaks down and stops. How does it work? Well as it spins in on itself it pulls the air through it with very little effort making it move along it axis. This is similar to a hard ball rolling across a flat floor.

The smoke ring is a type of vortex which pulls air into it and through it.

Our toroidal spinning electron model is like a smoke ring, an electromagnetic vortex.

It is interesting to imagine on the sub-atomic level that all fundamental particles act as spinning and convoluted electromagnetic standing waves, as vortices spinning at the speed of light and immersed in an energetic sea of an almost infinite supply of short lived particles and photons. It may be that the energetic space vacuum gets caught up in these power vortices and that there is a continuous exchange of quantum energy between the particle vortices and the space vacuum. This exchange of energy we see in the Lamb Shift of the quantum energy levels of the electron and the electron's observed jitter. The energy of the space vacuum may indeed be maintaining the spin of the electron around an average ground state energy level.

If indeed this model is a true picture of fundamental particles then matter particles may be acting just like mini-smoke rings sucking in and interacting with the space vacuum. It is probable that the 'infinite' supply of energy of the space vacuum maintains matter's very existence. Is it possible, therefore, that this flux of the space vacuum energy (virtual and real particles and photons) into a mass particle such as the electron is the mechanism for gravity.

In this model all matter particles are electromagnetic vortexes therefore gravity in this model is seen as a flux of the space vacuum energy into a mass (M). If we consider that the space vacuum energy of virtual and real particles is the matrix of space-time then any other mass (m) placed in this space-time energy flux will accelerate with the flux towards the mass (M) – just like a float in water accelerates towards a whirlpool around a sink plug hole.

Gravity may be an acceleration of invisible space-time vacuum energy into a mass.

CHAPTER 10
The Test of Space-Time
The acid tests of Newton's and Einstein's Gravity

Postulating that the fundamental nature of a matter particle is a toroidal electromagnetic standing wave spinning at c like a vortex, is there any evidence that our model acts as an influx of the space vacuum energy? If this gravity model is correct then a flux of the space vacuum energy into matter would have to satisfy all the criteria for gravity including *all* the effects observed by Newton's Law of Universal Gravity and Laws of Motion and Einstein's model of warped space-time. This model that matter is an energy sink for the energetic space vacuum has certain elegance.

If gravitational acceleration is an electromagnetic energy flux of space-time then an object in free fall (in the flux) should be at perfect rest (the test of an accelerating gravitational field) and this is certainly true for an object in fluidic flux: think of a floating object moving effortlessly with the flux of river water. It is equally being pulled and pushed and therefore feels no forces acting upon it. Another point, two masses by Newton's universal law of gravitation feel an equal and opposite attractive force, so would two sink masses: the flux of the space-vacuum energy would be into both masses and therefore they would be drawn together.

There is another point to be made here when talking about this energy flux: if we consider that the quantum foam of short lived virtual particles of space-time is a seething disturbance of the fabric of space-time then any flux of these particles into a mass is a flux of space-time itself.

In this model it is the very presence of the electromagnetic mass-energy of the space vacuum which gives the universe its dimensions of space (and time.) Think of a balloon filled with air – what is it without the air?

Using this model for gravity certain phenomena can be viewed in a different light, for instance, the way gravity acts in spiral galaxies. A spiral galaxy consists of billions of stars circling around its star dense galactic centre and accompanying super massive black hole, just like water being pulled down a plug hole in a vortex.

When the galaxy was formed by the condensation of a super cloud of primordial hydrogen in the early universe, collapsing under gravity, the matter gained kinetic energy resulting in a huge swirl around its centre. Recently astronomers have discovered that most galaxies including our own have super massive black holes at their heart which are pulling in neighbouring stars and absorbing them into their own mass with enormous release of energy (mostly in the form of x-rays). Stars at the centres of galaxies are being eaten alive by these black holes and as they orbit around the rotating galaxy they are slowly being pulled towards the galactic centre. With our model that gravity is a flux of the energetic space vacuum into a mass it doesn't take much imagination to see what's going on in the picture above.

This model of gravity makes sense of the concept of gravitational potential (stored) energy; for to lift a mass against gravity's space vacuum flux would require a force and once the mass was released it would simply accelerate with the flux.

Question: how much space-vacuum energy (E_v) does a mass like the Earth draw in from space per second?

Answer: this would be the energy influx power (Pv) of the sink mass, which is probably related to Gravitational potential Energy (GPE):

Gravitational potential energy is energy an object possesses because of its position in a gravitational field.

This will be an interesting estimate using the concept that the energy flux will do work on any unit mass placed in the flux and may not represent the *actual* energy flowing into a mass such as the Earth, nevertheless let us try and calculate the energy flux based on the sole concept of masses in freefall around the Earth for it may lead to some interesting conclusions.

We can assume for an object near the surface of the Earth where the gravitational acceleration 'g' is a constant at about 9.8 m/s². Since the zero of gravitational potential energy can be chosen at any point (like the choice of the zero of a coordinate system), the potential energy at a height h above that point is equal to the work which would be required to lift the object to that height with no net change in kinetic energy. Since the force required to lift it is equal to its weight (W = mg), it follows that the gravitational potential energy is equal to its weight times the height to which it is lifted:

$$GPE = mass \times g \times h$$

If a mass is released in free fall then the kinetic energy it *gains* (in the flux of the space vacuum) is equal to the gravitational potential energy lost.

Let us use unit mass (1kg) and unit time (1s):

Kinetic energy gained from the space vacuum (Ev)

= GPE lost per second

= kinetic energy gained per second

$$Ev = mgh = ½ \, mv^2$$

We know that after one second any mass dropped in free fall near the Earth's surface will gain a velocity (v) equal to its gravitational acceleration 'g' = 9.81 m/s.

Therefore:

$$Ev = ½ \times 1kg \times (9.81m/s)^2 = 48.118 \text{ Joules}$$

This tells us that any unit mass placed in free fall at any (and every) point around the Earth's surface will in one second gain 48.12 Joules of energy from the flux of the space vacuum Assuming this is a radial energy flux into the sink mass which follows an inverse square of the distance law, then:

$$Ev = \text{Power of mass}/4\pi r^2$$

Where r = the radius of the Earth = 6.357×10^6m

Then the total energy flux per second (Power) into the mass of the Earth:

$$Pv = Ev \times 4\pi r^2 = 48.118J \times 4\pi(6.357 \times 10^6 m)^2$$

$$Pv = 2.444 \times 10^{16} \text{ Watts}$$

In other words the Earth has a gravitational potential power of 2.444 x 10^{16} Watts. An interesting exercise will be to calculate the gravitational power of each molecule of the Earth. This will be an approximation because the Earth is composed of different density materials.

Let us assume the average density of the Earth is:

$$p_E = 5,515.3 \text{ kg/m3}$$

(5.5153 times more dense than water)

Therefore, assuming that the Earth 5.5 times denser than water which has a molar mass of 18g, then the molar mass we need is:

Average molar mass of Earth molecules

$$= 0.018kg \times 5.5153 = 0.09928kg$$

Since each mole of water contains Avogadro's number (N_A) of molecules, then the number of molecules in our 'water' Earth equals:

$$N = \text{Mass of Earth} \times N_A/0.09928kg$$

$$= 5.9742 \times 10^{24}kg \times 6.02 \times 10^{23}/0.018kg$$

$$N = 3.6226 \times 10^{49} \text{ molecules!}$$

Therefore the gravitational power of each 'molecule' in the Earth equals:

$$Pv_{mol} = Pv/N$$

$$= 2.444 \times 10^{16} \text{ Watts}/3.6226 \times 10^{49}$$

$$Pv_{mol} = 6.747 \times 10^{-34} \text{ Watts}$$

The gravitational potential power of the space vacuum energy flux into the Earth per molecule per second

Whether co-incidental or not the value 6.747 x 10^{-34} we have seen before in Planck's constant of 6.626 x 10^{-34}Js. This suggests that gravitational energy flux into a mass molecule comes close to the *magnitude* of Planck's constant.

Further Tests for the Model

Does this model of space vacuum energy flux stand up to the known characteristics and effects of gravitational fields? In its simplicity and elegance this model of gravity as a flux of the space vacuum energy makes sense of gravity and its effects – even a child can understand some object being carried along with a fluid flow. Imagine the Earth as a collection of a huge number of electromagnetic standing wave vortices each sucking in the space vacuum energy from every direction.

Place a mass in this flux and it accelerates towards the Earth (Test 1).

Question: why does the flux of space-time energy into a mass accelerate?

Answer: because in this model we would have a constant amount of energy fluxing into the mass which would follow an inverse square law.

Think of the volume of space energy as a giant balloon shrinking around the Earth accelerating in its collapse. In Newton's model of gravity the inverse square law of a gravitational field means that the gravitational field strength decreases by $1/r^2$ with distance from the centre of the mass. In Einstein's curvature of space-time it is the rate of change of velocity, acceleration, which decreases as an inverse square law. If we look at our model again and see gravity as a flux of space-time (space-vacuum energy) into a sink mass then the model of a shrinking sphere of space vacuum energy should give us an inverse square law for a point on the surface of the collapsing sphere. Let us put this model of a shrinking spherical volume to a mathematical test.

In unit time (Δt) the sphere of space-time would shrink by a constant fixed volume (ΔV). This volume of space-vacuum energy in a given time would be a constant for a given sink mass. As the space-vacuum energy falls into the mass then we can calculate the acceleration of a test mass in this flux as a change of (r), where:

$$\text{Volume of a sphere} = 4/3\pi r^3$$

$$r = \text{cube root } (3V/4\pi)$$

Taking unit values of (V) from 50 to 1 and calculating R for each volume we then calculate the change in R in unit time which gives us the distance travelled (x) by a point on the surface of the decreasing sphere. Since (x) is the distance travelled by the point in unit time (t), this is also its velocity (v). Acceleration is simply the change in velocity (dv) in unit time:

$$a = dt/t$$

The acceleration (a) is then plotted against the average value (R) for each change in volume. The results for these calculations are shown in the graphs of acceleration against R and $1/R^2$ below:

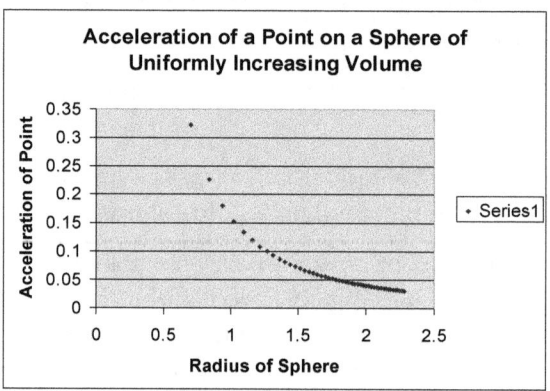

Acceleration of a collapsing sphere

The above curve of the graph clearly shows that the acceleration of a point on the surface of the shrinking sphere increases as the radius decreases and models the curvature of space-time.

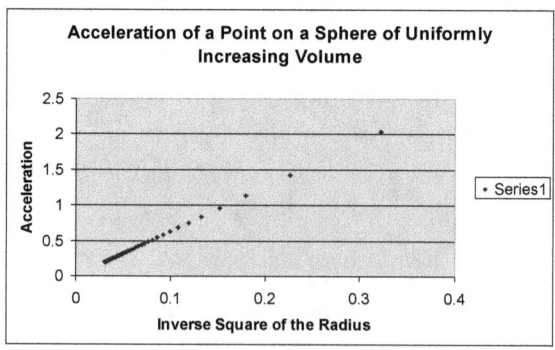

Acceleration follows an inverse square law

This graph demonstrates that this acceleration clearly follows an inverse square law where acceleration is proportional to $1/R^2$.

So the model of a flux of space-time vacuum energy into a mass definitely fits Newton's and Einstein's models of gravity.

Note: this model of a shrinking sphere producing an inverse square law for acceleration was proved correct by a friend of mine, Peter Burkhardt, who is a mathematician, in fact he made a flash animation of this model in action, which can be found at:

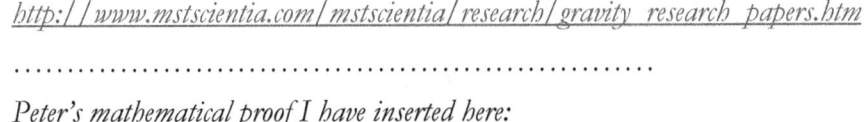

http://www.mstscientia.com/mstscientia/research/gravity_research_papers.htm

....

Peter's mathematical proof I have inserted here:

Shrinking Sphere

Problem:

Let us consider a sphere the volume of which is shrinking at a constant rate. Show that the change rate of the sphere's radius is inversely proportional to the square of the sphere's radius.

We introduce the following notations:

• V (t): The sphere's volume at time t.

• r(t): The sphere's radius at time t.

• c: The constant rate at which the volume is shrinking. Hence, we have to show that

$$dr(t)/dt = k \cdot 1/[r(t)]^2 ,$$

where k is a certain constant.

Solution:

Since the change rate of the sphere's volume is constant, we can write

$$dV(t)/dt = c. \qquad (1)$$

On the other hand, we can express the volume of a sphere in terms of its radius:

$$V (t) = 4/3 \pi \cdot [r(t)]^3. \qquad (2)$$

Using equation 2, we substitute V (t) in equation 1.

This yields

$$dV(t)/dt = d/dt (4/3 \pi \cdot [r(t)]^3) = c. \quad (3)$$

Applying the chain rule of differential calculus to the expression in the middle of equation 3, we find

$$d/dt \ (4/3 \ \pi \cdot [r(t)]^3) =$$

$$4/3 \ \pi \cdot 3 \cdot [r(t)]^2 \cdot dr(t)/dt =$$

$$4 \ \pi \cdot [r(t)]2 \cdot dr(t)/dt = c. \qquad (4)$$

Now, we solve equation 4 for dr(t)/dt and obtain:

$$dr(t)/dt = c/4\pi \cdot 1/[r(t)]^2, \qquad (5)$$

and we are done.

Peter Burkhadt

....

The final derived equation shows the radius shrinks at a rate which is proportional to $1/r^2$, in other words the collapsing sphere follows and inverse square law. The sphere accelerates as it shrinks. In Newton's gravity equation acceleration is also proportional to $1/r^2$:

$$a = F/m = GM \, . \, 1/r^2$$

In a gravitational field, what we see as free-fall is a mass accelerating with a collapsing sphere of space vacuum energy and therefore appears weightless (Test 2).

When the mass meets an obstacle it feels its own weight because it is stationary against the flux (Test 3) and therefore exerts a contact force against the obstacle (Test 4).

There is a flux of space-vacuum energy, and therefore acceleration, into both masses, which we see as the action and reaction forces of Newton's gravitational field theory (Test 5).

Any object moving at a tangential velocity to the Earth's surface will be caught by the space-time energy flux and will fall towards the mass in a parabolic path, at the right velocity it will orbit the mass (Test 6).

An object thrown upwards against the flux would decelerate upwards then accelerate downwards with the flux (Test 7).

This model, thus far, has passed all seven tests for the observable effects of gravity, but does the energy flux model fit with Einstein's view of curved space-time around a gravitating mass. Not only will the flux accelerate towards the sink mass, but, assuming a shrinking volume containing a constant amount of energy, then the space-vacuum energy density p_v will *increase* as the flux nears the mass, this will appear as space-time curvature – (Test 8).

Space-vacuum energy flux accelerates into the sink mass and as the result of a shrinking volume increases the energy density p_v. This appears as the curvature of space-time of a gravitational field

There are other effects of a gravitational field to test our model, these tests are 'tougher' and incorporate the other effects of Einstein's General Theory of Relativity.

1. Gravitational red-shifting of light (Test 9)

The wavelength of light increases (become 'redder') as it moves from a strong gravitational field to a weaker one.

This fits with our model of decreasing space-vacuum energy density as you move away from a mass, but why would its wavelength get longer when Pv is less. Simple refraction effects: the space-vacuum energy density is the medium in which the light travels:

More dense to less dense: wavelength increases and speed increases

Less dense to more dense: wavelength decrease and speed decreases.

The space-vacuum energy density changes produce the observable effects of refraction, including red-shift of light in a gravitational field.

Refraction of light as it leaves a dense region of the space-vacuum energy near a mass will produce a red shift: increasing wavelength and speed.

2. Time Dilation (Test 10)

A gravitational field distorts the dimensions of space-time. Clocks run slower in a strong gravitational field compared to an observer in a weak gravitational field: So would clocks run slower in a more dense space-vacuum energy? First remember what we said about refraction effects: the speed of light in a medium such as glass is much less than in a vacuum where it is (c). As the light enters a substance such as glass from a vacuum, travelling at c, the individual particles of light (photons) interact with atoms. The atoms act like antennae and absorb the photons and then after a specific interval (Δt) they transmit them.

This time delay gives light an apparent reduction in speed, *in between* the atoms there is still the vacuum of space where the speed of light is constant (c). When the light photons leave the glass medium they continue at high speed, at c.

So in the higher energy density space-vacuum near a sink mass, where there are more virtual and real particles and photons, the speed of light would be slower and since the speed of light governs time, by the speed of electromagnetic interactions, then time itself would run slower.

3. The Shapiro Delay (Test 11)

The Shapiro time delay effect, or gravitational time delay effect, is one of the four classic solar system tests of General relativity. Radar signals passing near a massive object take slightly longer to travel to a target and longer to return (as measured by the observer) than it would if the mass of the object were not present. This is easily explained by our model of increasing space-vacuum energy density near a sink mass, this is the effect of refraction: the reduction of the speed of light though the medium of the space-time energy density.

4. Bending of light (Test 12)

When light passes through a gravitational field it follows a curved path.

Gravity bends light

Again this is a refraction effect explained by the speed change of light as it passes through the increasing energy density of the space-vacuum near the sink mass.

6. Gravity is Acceleration – Warped Space-time (Test 13 and 14)

In Einstein's view of gravity, gravity is not a force but acceleration as a result of the curvature of space and time. In a gravitational field all objects fall (accelerate) at the same rate (Test 13), their acceleration is independent of mass (Test14). This is true in the accelerating flux of the space-vacuum energy into a sink mass – all objects in this flux would accelerate (flux) with space-time at the same rate. The larger the sink mass the bigger its inertia, but also the more space-vacuum energy flux into it. The model of space-vacuum energy flux into a mass passes all fourteen tests for the observable effects of gravitational fields, but doe it pass the next and more stringent test of Einstein's Theory of Special relativity?

Einstein's Special Theory of Relativity

How does our model of the space vacuum match up to Einstein's special theory of relativity? Here the speed of light is a constant and therefore time and space are variable. When objects are travelling at near light speeds, time slows down. At the speed of light itself time theoretically stops. To answer this question properly we must re-examine the speed effects of light refraction:

This universal constant (c) of Einstein's world as you know is fundamental to physics and as we have already seen when light travels through a medium there is an apparent reduction in its speed due to absorption and re-emission of light photons by atoms which takes a short interval of time (Δt). In between the atoms light travels at c and so is still a true constant. But does light travel at a finite and fixed speed (c) in the space vacuum? The fact, that theoretically nothing can travel faster than the speed of light, appears a s a physical barrier to those who dream of someday travelling to the stars at superluminal speeds. This reminds me of the time when test pilots were trying to break the speed of sound barrier – go supersonic, which of course they did, but the speed barrier of the universe, the speed of light, seems unbreakable, but is it?

In answer to this fundamental question we return to the space vacuum energy density which by the absorptions (Δt) which slows down the speed of light and hence time: What follows is a pretty remarkable postulate about the speed of light: in the true nature of the universe: Let's imagine the impossible or at least the improbable:

The speed of light is infinite

Is this possibly true, however incredible this statement may seem? How can the speed of light possibly be infinite when it is a visible and measurable constant in the vacuum of space? But, remember space is not empty. Through what we call empty space there is an enormous amount of energy, the energy of the space vacuum, a seething sea of photons and short lived mass energy particles.

If light is slowed down in a matter medium by the Δt absorptions of particles then it must be slowed down in the space vacuum in the same way.

These absorptions in the seething sea of the space vacuum give light its measurable finite speed c.

The space-vacuum energy density of the universe dictates the speed of light (c) and hence the speed of time (1 sec per second). In a reduced space-vacuum energy density the speed of light would increase by refraction to faster than c and hence time would run faster (relatively). In this case if you reduce the space-vacuum energy density to zero the speed of light would approach infinity. In a true vacuum therefore the speed of light is infinite, but in a true vacuum with absolutely no mass or energy at all, space and time, and therefore the universe could not exist. Without the space-vacuum energy time would not exist. It would be meaningless if all electromagnetic reactions are truly instantaneous. The creation of space-time and mass energy limited the speed of light in the universe and also created time as we know it. Time started with the Big Bang and the formation of this universe.

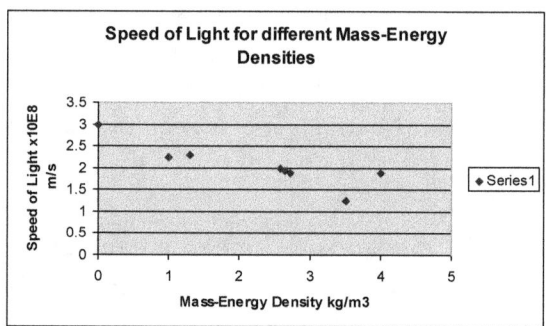

This graph showing how the speed of light increases as the mass-energy density of matter decreases.

At 'zero' density, the current density of the space vacuum, the speed of light is c, from the slope of the graph it can be deduced if the space vacuum energy density decreases to zero then the speed of light will approach infinity!

Entanglement

Einstein's speed barrier of c in the vacuum remained intact for many years. In his universe no information or electromagnetic effects between particles, or gravitational effects between masses could happen at speeds greater than the speed of light, but recent developments in physics have revealed, with experimental proof, that communication of information between particles *can* actually happen at speeds greater than the speed of light, in fact particles can interact over large distances instantaneously! This is called entanglement. Entanglement is a term used in quantum theory to describe the way that particles of energy/matter can become *correlated* to predictably interact with each other regardless of how far apart they are.

Particles, such as photons and electrons, that have interacted with each other retain a type of connection and can be entangled with each other in pairs, in the process known as correlation. Knowing the spin state of one entangled particle - whether the direction of the spin is up or down - allows one to know that the spin of its mate is in the opposite direction. Even more amazing is the knowledge that, due to the phenomenon of superposition, the measured particle has no single spin direction before being measured, but is simultaneously in both a spin-up and spin-down state. The spin state of the particle being measured is decided at the time of measurement and communicated to the correlated particle, which <u>simultaneously</u> assumes the opposite spin direction to that of the measured particle.

It seems that particles which are separated by incredible distances interact with each other instantaneously, in a communication that is not limited to the speed of light. No matter how great the distance between the correlated particles, they will remain entangled as long as they are isolated. Entanglement is a real phenomenon, Einstein called it 'spooky action at a distance', which has been demonstrated repeatedly through experimentation.

The mechanism behind it has not been fully explained by any theory in physics. Entanglement therefore shows that somehow there is an instantaneous mechanism in the universe for communication/interaction between particles. This may support the theory that the speed of light is infinite. In our model for the space-vacuum energy where the speed of light is limited only by the space-time energy density, in the true vacuum between the short lived mass-energy particles light communication may travel at infinite speed. Hence particles can be influenced electromagnetically by each other – instantly! Recent evidence for faster than light communication between particles:

String Net Liquid of the Space Vacuum

An article published by the New Scientist magazine (March 17 2007) refers to experimental evidence for the string net liquid nature of the space vacuum. String net liquid theory suggests that the space vacuum is a tangible fluid from which fundamental particles arise: 'particles arise from the deeper structure of the non empty vacuum of space-time. The net strings vibrate according to Maxwell's equations which describe the behaviour of light. Elementary particles form at the ends of the strings. The electrons in a crystal net behave as if entangled (communicate with each other instantaneously) even though being separated by great distances. From this, the researchers made another leap: That maybe the vacuum of our whole universe is a string-net liquid. The new theory claims to provide a unified explanation of how both light and matter arise. According to this new theory backed up by experimental evidence of the way electrons behave in a crystal of herbertsmithite the space vacuum is a fluid in which entangled particles exist and allowing instantaneous effects. In our model of gravity this space-vacuum energy density is a 'fluid' which flows into a sink mass.

Special Theory of Relativity

Newton's laws of motion give us a complete description of the behaviour moving objects at low speeds. The laws are different when the speeds of particles approach the speed of light (c). Einstein's Special Theory of Relativity describes the motion of particles moving at close to the speed of light. In fact, it gives the correct laws of motion for any particle. This doesn't mean Newton was wrong; his equations are contained within the relativistic equations. Newton's 'laws' provide a very good approximate form, valid when v is much less than c. For particles moving at slow speeds (very much less than the speed of light), the differences between Einstein's laws of motion and those derived by Newton are tiny. That's why relativity doesn't play a large role in everyday life. Einstein's theory supersedes Newton's, but

Newton's theory provides a very good approximation for objects moving at everyday speeds.

Einstein's theory is now very well established as the correct description of motion of relativistic objects, which are those travelling at a significant fraction of the speed of light. Because most of us have little experience with objects moving at speeds near the speed of light, Einstein's predictions may seem strange. However, many years of high energy physics experiments have thoroughly tested Einstein's theory and shown that it fits all results to date. Einstein's theory of special relativity results from two statements - the two basic postulates of special relativity:

1. The speed of light is the same for all observers, no matter what their relative speeds.

2. The laws of physics are the same in any inertial (that is, non-accelerated) frame of reference. This means that the laws of physics observed by a hypothetical observer travelling with a relativistic particle must be the same as those observed by an observer who is stationary in the laboratory.

Given these two statements, Einstein showed how definitions of momentum and energy must be refined and how quantities such as length and time must change from one observer to another in order to get consistent results for physical quantities such as particle half-life. To decide whether his postulates are a correct theory of nature, physicists test whether the predictions of Einstein's theory match observations. Indeed many such tests have been made - and the answers Einstein gave are right every time.

Gamma(γ)

The measurable effects of relativity are based on gamma (γ). Gamma depends only on the speed of a particle and is always larger than 1. By definition:

$$\gamma = 1/\text{SQR}[1 - (v^2/c^2)]$$

Where: c is the speed of light and v is the speed of the object in question

For example, when an electron has travelled ten feet along a particle accelerator it has a speed of 0.99c, and the value of gamma at that speed is 7.09. When the electron reaches the end of the accelerator, its speed is 0.99999999995c where gamma equals 100,000. What do these gamma values tell us about the relativistic effects? When gamma is 7.09 it means that time slows down by this factor and lengths contract by this factor. Notice that when the speed of the object is very much less than the speed of light ($v << c$), gamma is approximately equal to 1. This is a non-relativistic situation (Newtonian).

Length Contraction and Time Dilation

One of the strangest parts of special relativity is the conclusion that two observers who are moving relative to one another, will get different measurements of the length of a particular object or the time that passes between two events. Consider two observers, each in a space-ship laboratory containing clocks and meter sticks. The space ships are moving relative to each other at a speed close to the speed of light. Using Einstein's theory:

Each observer will see the meter stick of the other as shorter than their own, by the same factor gamma. This is called **length contraction**.

Each observer will see the clocks in the other laboratory as ticking more slowly than the clocks in his/her own, by a factor gamma. This is called **time dilation**.

So at 99% the speed of light the observer sees that time for the particle has slowed down by a factor of 7.09! Time for the passenger runs seven times slower than for the observer! At 99.999999995% the speed of light time is dilated by 100,000 times. Let us examine what happens in our model for the space vacuum as we approach near light speeds.

In this test we will use a laboratory of fixed length (L). Inside the laboratory at one end we have a laser and at the other end a stop clock which times (t) the passage of the beam across the room through a super cooled (2.7K) tube which has a true vacuum. The only thing inside the tube which the laser travels through is the space vacuum energy at density (Pv).

The speed of the laser light is thus calculated from:

$$\text{Speed of light} = L/t$$

Stationary Laboratory

The laser is turned on and the clock starts running automatically. As the beam traverses the space-vacuum encountering the seething sea of particles there will be numerous absorptions and re-emissions, each one having time Δt. In our model we are postulating that in a true vacuum free of energy density the speed of light is infinite. To simplify the picture (and the maths) let us also assume that between the space vacuum particles there is zero energy and zero electromagnetic influence on the speed of light. Therefore when the beam reaches the other end and stops the stop clock the time to traverse the room will be the sum of all the little Δts $= \Sigma \Delta t$.

Therefore for our stationary laboratory the speed of light is seen as (c):

$$c = L / \Sigma \Delta t$$

So far all is well with the universe.

Now let us put the same laboratory in a spaceship and start moving through 'empty' space at near light speeds:

Laboratory at Near Light Speeds

Our laboratory is traversing through the space vacuum energy where the mass energy particles easily pass through the real matter of the ship. The space vacuum energy is like a wind through the transparent ship – this effectively increases the energy density in the laboratory. While the laser beam is traversing the laboratory is passing through many more virtual pairs of particles, therefore there will be a much larger number of Δts. So $\Sigma \Delta t$ will be larger. The beam will take longer to pass across the laboratory. And since:

$$v = L / \Sigma \Delta t$$

The result of this is that the speed of the light traversing the laboratory will decrease. It will take longer to reach the end and since the speed of light governs time, if the speed of light in our laboratory is decreased then all electromagnetic interactions will slow down, time runs slower. Would we notice the difference, because our brain's chemical reactions are slower, our thinking slows down by the same exact amount. Time would be running as normal for us in the spaceship, but for a stationary observer he would see time in our spaceship running slower – a relativistic effect. If our spaceship was actually travelling at the speed of light the space vacuum energy would be fluxing though the laboratory at the same speed, the speed of light in this case would appear zero. Time stops.

The effects on length contraction are the same: assuming the space vacuum energy of space time which gives space and matter its length gets squashed. The picture of warped (curved) space-time in Einstein's universe is difficult to imagine – what exactly is bent if space devoid of mass-energy has no substance? In our model of the universe space-time *is* the space vacuum energy, it is the presence of this probably infinite source of mass-energy which gives space its dimensions and dictates by its energy density the speed of light and therefore gives us universe time. In our current universe the space-vacuum energy density (ρ_v) is such that the speed of light through it is (c) = 2.99792458×10^8m/s and real time (T) as a consequence runs at a rate of one second per second. Real time T is from the observer in the frame of reference of the energy density   Increase the energy density ρv then c decreases and time t increases i.e. one second is longer than one second. Therefore time is proportional to the speed of light and inversely proportional to the space-vacuum energy density:

$$P_v \; \alpha \; 1/c \; \alpha \; 1/t$$

Doubling the energy density means that the speed of light is halved (c/2) and also time runs half at half its normal rate (t/2). Reactions that would normally take one second now take twice as long = 2 seconds/second, but the observer in the energy frame would not notice because all reactions are slowed down even our thinking. This gives us an **absolute time formula:**

$$T = k_\gamma \rho_v c$$

Where gamma (k_γ) is a constant the value of which can only be determined if the energy density is known. From the formula real time (T) for the observer is a constant because doubling ρv halves c! Only an observer outside the energy frame would notice the difference in c and T. This is special relativity.

Gravity the Force which Cannot be Shielded

Another test for our model is the unique property of gravity: gravity acts through all objects, unlike the forces of electro-magnetism which can be shielded. This is a test for a theory of gravity and as we have seen the flux of the space vacuum will permeate through matter. If our theory is correct then as you sit in your chair reading this book space-time in the form of space vacuum energy is fluxing though you to the Earth and into you towards every atom particle in your body.

Space vacuum energy particles are virtual and pervade the space between matter and therefore can flux through matter

It is only when the space vacuum *accelerates* through us does it exert a drag force which we see as weight. It seems our model for gravity stands up to the rigid tests of Einstein's General Theory of Relativity, not only that, it seems to makes sense of these observable and measurable effects. If the model is correct:

It is the flux of the space vacuum energy density (short lived virtual particles and photons) into a mass particle such as the electron which we see as gravity. In this model all matter particles are electromagnetic vortexes therefore gravity in this model is seen as a flux of the space vacuum energy into a mass (M). If we consider that the space vacuum energy of virtual and real particles is the matrix of space-time any other mass (m) placed in this space-time energy flux will accelerate with the flux towards the mass (M). Near the mass the space vacuum energy density increases giving us the observed effects of warped space time, including time dilation and red shift.

Newton's Laws of Motion and the Space-vacuum Energy Density

At school we learned Newton's three laws of motion:

First Law:

An object at rest will remain at rest unless acted upon by an external and unbalanced force. An object in motion will remain in constant motion unless acted upon by an external and unbalanced force.

This first law means that an object should keep moving at a constant speed and direction unless acted on by a force and seems contradictory to our experience, for we if we push a toy car it will quickly slow down and change direction – they definitely never move straight. And as we are taught it is because of the external forces of friction of the wheels with the ground which causes the trolley to lose energy and stop. Thankfully in space, a traditional vacuum, well away from a gravitating mass, there is no friction and therefore nothing to stop a moving object or change its direction. This principle is used in space travel, once a rocket has accelerated up to speed the engines are turned off and the rocket cruises at constant speed and in a straight line according to Newton's first law of motion. Before it can do this it must escape the Earth's gravitational field – it must reach escape velocity, for the Earth this is 11km/s. At this speed it has enough kinetic energy to escape the Earth's gravity completely. It is like throwing a ball into the air, give it enough velocity and it will never come down.

One thing which is difficult to understand is why when away from a gravitational field a pushed object keeps moving? The answer I give my students is: that there is nothing to stop it moving, there are no frictional forces, there is no drag in empty space, but as we have seen space is not empty but filled with an enormous amount of space-vacuum mass-energy. Some calculations show this density of the space-vacuum energy to be equal to atomic nuclear densities. Nuclear density is the density of the nucleus of an atom, averaging about 10^{18} kgm^{-3}, the density of steel for instance is only 8 x 10^3kg/m^3. So how can an object move at constant speed through such a dense space-vacuum medium? The answer to this lies in the nature of matter – it must be transparent to the space-vacuum energy density when moving at a constant speed. The quantum vacuum fluctuations of space-time do not exert a drag force when an object is moving through it at constant speed. Matter is *transparent* to the quantum particles of the space-vacuum energy density, because matter particles have a much *bigger wavelength* than the extremely dense and small wavelength particles of the space-vacuum energy density. This is like light (long wavelength) travelling through solid glass (short wavelength atoms).

Newton's Second Law:

The rate of change of momentum of a body is proportional to the resultant force acting on the body and is in the same direction.

This means that to change the constant motion of an object requires external forces acting on the object; the result is the object accelerates. Note: the acceleration can be a change in speed or direction. So why does our free moving object in space travelling at a constant speed where it experiences no drag from the space vacuum energy density, require a force to accelerate it? The answer must be that:

Acceleration (a) through the space vacuum energy density produces a drag (F) on a mass.

This drag (F) must be directly proportional to the acceleration (a) and the interacting mass (m) of the object moving through the space vacuum medium. Hence: $F = ma$.

Question: why does a mass experience a drag from the space vacuum medium only when it is accelerating and not when moving at a constant speed?

Answer: when accelerating there are electromagnetic interactions between the fields of matter particles and space-vacuum particles.

At constant speed there is an equilibrium state between the electromagnetic fields of the matter particles and the space vacuum particles. To change this equilibrium state requires a force to give the mass more energy so that it can undergo more interactions and hence reach a new equilibrium state of electromagnetic interactions of attractions and repulsions.

It can be stated that in accelerated motion there are changes in Doppler shift between the particles and this results in the mass experiencing a drag force from the space vacuum

This explains why a mass has inertia:

Inertia is the property of an object to remain at constant velocity unless acted upon by an outside force. The principle of inertia is one of the fundamental principles of classical physics which are used to describe the motion of matter and how it is affected by applied forces. Inertia of a mass is the reason an object keeps moving and why it is difficult to change a mass's inertia by slowing it down, speeding it up or changing its direction. We have all experienced inertia when standing in a bus which suddenly speeds up or slows down, in the first we sway backwards and in the latter we sway forwards because our body's inertia wants to carry on at the same speed – it resists change.

This raises a very interesting point concerning gravity and perhaps a corner stone of the model.

If an acceleration of a mass through the space vacuum energy produces a drag force on the mass, then an acceleration of the space-vacuum energy through a mass must also produce a drag force on the mass!

This must be the mechanism of gravity – an *acceleration* of the space vacuum energy (space-time) through a mass which produces a *drag force* which we see as gravity. Newton's Laws of motion prove that gravity is an *acceleration* of the space vacuum energy through a mass.

Further Notes on Newton's Laws being a Corner Stone for this Gravity Model

It is interesting to speculate as to what gives space its optical-electromagnetic energy density properties. It is not difficult to suppose that what we see as empty space is not empty at all, but whatever fills space must be a very elusive substance considering that moving a mass through the substance of space has very little effect, matter seems to move through this ethereal substance with little difficulty.

This drag force is described by Newton's 2nd Law of Motion:

$$F = ma$$

The drag force (F) is proportional to the mass (m) and the acceleration (a) of the mass particle through the medium of the space vacuum. In Einstein's mass equivalence principle there is no difference between inertial mass (m) and gravitational mass (m). We measure this gravitational mass by its weight (w) in a gravitational field where the gravitational acceleration (a) equals g: w =mg

This mass equivalence gives us the insight we need in exactly how a gravitational field behaves, if accelerating a mass through the substance of space exerts a drag force (F) on a mass, then the weight that a stationary mass particle experiences in a gravitational field must be the result of the substance of space accelerating though it!

$$F = ma \text{ is equivalent to } W = mg, \text{ where } m = m$$

It can be concluded, therefore, that a gravitational field is an acceleration of the electromagnetic energy density substance of space into a gravitating mass. A mass placed in this flux will experience a drag force which will accelerate it into free fall, once falling with the accelerating substance of space it is at perfect rest with the space around it. Any object placed it its path will cause the drag force of space to give it its known weight.

A final thought on our model of gravity: we have said before that gravity unlike the electric and magnetic forces is always an attractive force; this can be explained by our model because gravity is an influx of the space-vacuum into a mass.

CHAPTER 11
The Cyclic Nature of the Universal
Space Vacuum Energy

*Massive black holes at the heart of galaxies pump energy into the universe.
Neutron stars and pulsars, super massive black holes, at the centres of galaxies are
sucking in mass by eating stars and then pumping this mass energy in the form of
gamma rays and high speed electrons back out into the universe.*

Returning to our model of matter particles as electromagnetic standing
waves absorbing energy from the space vacuum, one question remains
unanswered. Where does this energy go? By the law of the
conservation of energy it must end up somewhere? If a matter particles
absorbs a photon of energy raising it above the ground state it will very
quickly transmit the energy as a single photon. There is one effect
which proves the existence of the space vacuum energy and this is the
Lamb Shift:

Lamb Shift

*Lamb shift is the slight shift in the energy and therefore shift in wavelength and
frequency, of an electron bound to a nucleus due to energy fluctuations in the space
vacuum. This shift in the electrons energy level is evidence for the presence of virtual
pairs of particles, such as the electron and positron or virtual photons. The electron
in its 'orbit' around the nucleus wobbles because of these energy fluctuations – this
effect which is seen and measured is called Zitterbewegung motion.*

Thinking of the Lamb Shift in terms of absorbed energy by the
electron we calculated this to be equal to: 5.4154×10^{-24}J.

*The electron after absorption of a quantum of energy is in an excited state and will
soon return to the ground state in a very short time by emitting a photon of
equivalent energy.*

So the Lamb Shift energy absorbed by the electron will quickly be
radiated as a photon of light. This describes where the space vacuum
energy goes after absorption it returns to the space vacuum as
radiation. We have an uptake of energy by matter particles immersed

in the space vacuum's seething energetic sea of virtual particles and after absorption this energy is radiated back into 'empty space.'

There is a continuous exchange of energy between the electron and itself and the sea of virtual particles and photons:

1. Emission and re-absorption of a virtual photon.

2. Emission and re-absorption of a virtual electron-positron pair. These interactions involve a single loop of virtual particles and are the first-order contribution to the Lamb shift. More complex interactions are possible, in which two or more loops of virtual particles are generated:

Examples of a two-loop contribution to the Lamb shift.

4. An electron emits one photon and then, soon after, another, so that the two overlap in time.

4. A virtual electron–positron pair is emitted, the virtual electron emits a photon that is reabsorbed.

We calculated that the energy-frequency of the Lamb Shift for an electron and found that it was equivalent to a peak wavelength of: 3.671cm. Therefore matter must be radiating energy at this wavelength back to the space vacuum – the wavelength of this radiation is in the microwave region, which falls within the spectrum of cosmic background radiation. We also calculated that this radiation was equivalent to a black body radiation temperature of: $T = 0.0789K$ – just below the peak temperature of empty space caused by Cosmic Background Radiation.

The energy Lamb Shift of electrons is equivalent to a black body temperature 0.0789K and a wavelength of 3.671cm – the microwave part of the spectrum

This equivalence of wavelength between Lamb Shift photons and Cosmic Background Radiation may be coincidental, but it's interesting to note that the interaction of matter particles immersed in the space vacuum jitters around the same energy levels. A question comes to mind in our model for gravity:

Is the mechanism for gravity, which is an absorption and emission of the space vacuum energy, which possibly results in an acceleration of the space-time matrix into matter responsible for CMBR?

The wavelength for lamb shift radiation falls well within the range for Cosmic Background Microwave Radiation and therefore the two may very well be linked. If this is true, then it is difficult to think of an analogy for this model, but here is one which may suffice:

The Mass-Energy Gravity Bucket

In this analogy the bucket is a mass particle and the flux of water into it is the space vacuum energy, water is pumped under pressure (space vacuum energy pressure) into the bucket through a pipe at the bottom, the buck fills up (ground state = full). A cistern float valve in the bucket maintains the bucket's full level, but invariably, over spill and splashes (Lamb Shift Radiation) return back to a source tank of water (the space vacuum). If indeed a particle takes energy from the space vacuum to maintain its ground state then we will have to simulate this by putting a hole in the bottom off the bucket!

It is easy to imagine a cycle of energy between matter and the 'infinite' energy of the space vacuum, this energy into matter, in the form of absorbed virtual particle pairs and photons, sustains matter at ground state energy levels and results in emissions at various microwave wavelengths, thus giving rise to CMBR which maintains the universe at an average temperature of 2.725K. This resulting influx of energy into matter is what we know as gravity. Any mass particle placed in this accelerating (a = g) influx will experience a drag force which gives rise to free fall and weight:

$$Drag\ Force = F = W = mg.$$

If massive objects are indeed converting space vacuum energy into gravitational microwave radiation, then in principle this radiation should heat up the matter in the body. This may explain why planets such as Jupiter emit more energy than they absorb from the sun. Jupiter emits 1.6 times more energy than it absorbs. Some of this energy can be explained by other means such as the presence of radioactive isotopes in its core, but Jupiter's main energy generator has remained a mystery for some time. The Earth too has a hot core and remains volcanically active – is the flux of space vacuum energy into the Earth responsible partly responsible for the Earth's internal temperatures remaining so high?

If space vacuum energy influx and radiation at microwave wavelengths is true then we have an energy cycle from the space vacuum into a mass and back out into space.

If the above mechanisms are correct for gravity then we can see why it is such an intrinsic 'force' in our universe: gravity is a mass- energy influx into (driven by the space vacuum energy pressure) into atomic mass particle vortices, which radiate the energy back into the zero point energy of the universe by virtual pair and photon exchange and by a spectrum of microwave radiation.

Note: In a conversation with my son on this possible cyclic energy nature for gravity, he suggested that these phenomena must be part of a vast universal energy cycle, where planets and stars take in energy from the space vacuum and power houses such as pulsars help maintain the energy of the space vacuum – where they convert matter to energetic particles and high energy gamma rays.

The Conservation of Space-time Energy

The concept, which states the space vacuum energy in the universe, is being continually being topped up by various phenomenon means that these powerful phenomena create and maintain a space vacuum energy pressure. It is this pressure (Pv) of the energy density (ρv) which maintains the flux into the masses of stars and planets. The vacuum pressure inside a spinning matter particle vortex will be lower creating a pressure difference which maintains the influx: perfect conditions for a vortex to be maintained. This pressure difference we also see in the Casmir effect.

Question: what would be the magnitude of the Casmir Pressure Pc for an electron?

Answer: Assuming A = area of sphere of electron – only certain resonant frequencies aloud inside it orbital spin therefore a reduced pressure inside:

Casmir effect force from the space vacuum: $F = 3.67 \times 10^{21} N$

Casmir effect pressure from the space vacuum: $Pc = 5.536 \times 10^{45}$ N/m^2

This is an enormous space vacuum energy pressure difference between the inside and outside of the electron sphere of spin. As in all cycles equilibrium is reached which we see in part as the temperature of empty space: the cosmic microwave background radiation at a temperature of 2.7K.

History of Gravity

In the early universe when matter particles first condensed out of light energy: The universe was much smaller and at extremely high temperature. This small universe was filled with a dense seething sea of high energy electromagnetic gamma wave photons – light energy. These free gamma photons interacted and formed standing wave particles of matter and anti-matter pairs which quickly annihilated each other back into free gamma rays. As the universe expanded and cooled the fundamental matter particles formed atoms, mostly hydrogen. At this time the space vacuum energy density could have been extremely high and therefore the speed of light would have been much reduced – as a consequence time would run slower. With the expansion of the universe the space vacuum energy density must have decreased to its constant level and temperature 2.7K, in this scenario, therefore light speed would have increased to c and time speeded up to its present 'rate' of one second per second. Antimatter particles which have a shorter decay time than matter particles left the universe filled with matter only.

Once real matter had formed as stable protons and electrons the universe was filled with mostly elemental hydrogen atoms. Under gravity these clouds collapsed to form stars. It was with the condensation of free wave light energy to standing wave particles that gravity came into existence for without the particle vortexes the pressure of the space vacuum energy had no mechanism, once particles were formed the influx of ether energy into mass particles started. Without matter there is no energy influx and therefore no gravity. Note: the absence of gravity in an early universe, devoid of matter particles, would have caused rapid inflation.

The Power Houses of the Universe

1. Stars like our Sun

Earth's Sun is a medium-sized star which lies on the main sequence (life time of hydrogen burning) with 90% of the known stars. It has an effective surface temperature is 5780 K, putting it in spectral class G2 (Yellow). Its mass is 1.989×10^{30} kg and its mean radius is 6.96×10^8 meters. The mass of the sun is over 99.8% of the mass of the entire known solar system. Our own sun like any other star is a power house pumping mass-energy back into the universe in the form of electromagnetic radiation and charged particles:

Electromagnetic radiation: from short wavelength/high energy x-rays to long wavelength-low energy radio waves. Due to the nuclear fusion processes in the core of our sun it radiates energy at the enormous rate of 3.85×10^{26} Watts (Joules per second).

Charged particles: the sun emits thousands of tonnes of charged particles into space at high velocities; the particles include protons and electrons travelling towards Earth as the Solar wind.

Neutrinos: During the nuclear fusion of hydrogen into helium the core emits billions of high energy neutrinos per second into space. These particles have no charge and almost zero mass and are travelling at almost light speed, they therefore can travel straight though matter without interacting at all, most of the solar stream of neutrinos from the sun will travel straight through the Earth and through you.

Gravity causes enormous pressures and temperatures in the core which drives the sun's nuclear energy producing processes. In our model for gravity, the influx of space vacuum energy into the mass of the sun is the power source which helps drive the stellar processes. The space vacuum energy influx into mass may be powering the nuclear processes of stars where mass is directly converted energy by E $= mc^2$. This energy cycle of the stars means that the universe is self sustaining. Stellar radiation of mass-energy back into the space vacuum helps maintain the space vacuum energy (and space density ρv at a constant equilibrium level. The current energy level density must have taken billions of years to reach its current equilibrium.

Supernovae and the formation of Neutron Stars and Black Holes

2. Supernovae

A supernova is a stellar explosion which produces an extremely luminous object that is initially made of plasma—an ionized state of matter. A supernova may briefly out-shine its entire host galaxy before fading from view over several weeks or months. During this brief period of time, the supernova radiates as much energy as the Sun would emit over about 10 billion years. The explosion expels much or all of a star's material at a velocity up to a tenth the speed of light, driving a shock wave into the surrounding interstellar gas. This shock sweeps up an expanding shell of gas and dust called a supernova remnant.

There are several types of supernovae and at least two possible routes to their formation. A massive star may cease to generate energy from the nuclear fusion of atoms in its core, and collapse under the force of its own gravity to form a neutron star or black hole. Alternatively, a white dwarf star may accumulate material from a companion star (either through accretion or a collision) until it nears the Chandrasekhar limit of roughly 1.44 times the mass of the Sun, at which point it undergoes runaway nuclear fusion in its interior, completely disrupting the star. This second type of supernova is distinct from a surface thermonuclear explosion on a white dwarf, which is called a nova. Solitary stars with a mass below approximately 8 solar masses, such as the Sun itself, evolve into white dwarfs without ever becoming supernovae.

A supernova is a relatively rare event that occurs, on average, only once every 50 years in a galaxy the size of the Milky Way. However, supernovas play a significant role in enriching the interstellar medium with heavy elements, and the expanding shock waves from these explosions can trigger the formation of new stars – all part of the mass-energy cycle of the universe.

3. Neutron Stars

A neutron star is one of the few possible endpoints of stellar evolution. A neutron star is formed from the collapsed remnant of a massive star after a Type II, Type Ib, or Type Ic supernova. Neutron stars are mighty power houses for the cycle of mass-energy in the universe: Not all neutron stars are destined to lead a life of isolation. Some of them are born in binaries that survive the supernova explosion that created the neutron star, and in dense stellar regions such as globular clusters some neutron stars may be able to capture companions. In either case, mass may be transferred from the companion to the neutron star.

If the companion star has less than the mass of our Sun, the mass transfer occurs via Roche lobe overflux. If part of the companion star's envelope is close enough to the neutron star, the neutron star's gravitational attraction on that part of the envelope is greater than the companion star's attraction, with the result that the gas in the envelope falls onto the neutron star. However, since the neutron star is tiny, astronomically speaking, the gas has too much angular momentum to fall on the star directly and therefore orbits around the star in an accretion disk. Within the disk, magnetic or viscous forces operate to allow the gas in the disk to drift in slowly as it orbits, and to eventually reach the stellar surface. If the magnetic field at the neutron star's surface exceeds about 10^8 G, then before the gas gets to the stellar surface the field can couple strongly to the matter and force it to flux along field lines to the magnetic poles. The friction of the gas with itself as it spirals in towards the neutron star heats the gas to millions of degrees, and causes it to emit X-rays.

By Stefan's Law the wavelength of the emitted radiation energy is:

$$\lambda max = 0.002898/1 \times 10^7 K = 2.90 \times 10^{-10} m$$

This falls within the range of x-rays (10^{-8}m to 10^{-11}m), more energetic gamma rays are (10^{-11} to 10^{-15}m). The neutron star is a powerful mass to radiation energy converter eating up the mass of neighbouring stars and sending the energy into the space vacuum. This process helps maintain the energy density of the vacuum of space. Neutron stars are not the most powerful mass energy converters; there is one type in a class of its own when it comes to converting mass into energy:

4. Black Holes

A black hole is an object with a gravitational field so powerful that even electromagnetic radiation (such as light) cannot escape its pull.

Formation of stellar-mass black holes

Stellar-mass black holes are formed in two ways:

1. The gravitational collapse of a star.

2. Collisions between neutron stars.

Although neutron stars are fairly common, collisions appear to be very rare. Neutron stars are also formed by gravitational collapse, which is therefore ultimately responsible for all stellar-mass black holes. Stars undergo gravitational collapse when they can no longer resist the pressure of their own gravity. This usually occurs either because a star has too little 'fuel' left to maintain its temperature, or because a star which would have been stable receives a lot of extra matter in a way which does not raise its core temperature. In either case the star's temperature is no longer high enough to prevent it from collapsing under its own weight. The collapse transforms the matter in the star's core into a denser state which forms one of the types of compact star. Which type of compact star is formed depends on the mass of the remnant, i.e. of the matter left to be compressed after the supernova triggered by the collapse has blown away the outer layers.

Only the largest remnants, those exceeding 5 solar masses, generate enough pressure to produce black holes, because singularities are the most radically transformed state of matter known to physics (if you can still call it matter) and the force which resists this level of compression, neutron degeneracy pressure, is extremely strong. Remnants exceeding 5 solar masses are produced by stars which were over 20 solar masses before the collapse (the rest of the mass is usually blown into space by the supernova triggered by the collapse).

In stars which are too large to form white dwarfs, the collapse releases energy which usually produces a supernova, blowing the star's outer layers into space so that they form a spectacular nebula.

Formation of larger black holes

There are two main ways in which black holes of larger than stellar mass can be formed:

1. Stellar-mass black holes may act as 'seeds' which grow by absorbing mass from interstellar gas and dust, stars and planets or smaller black holes.

2. Star clusters of large total mass may be merged into single bodies by their members' gravitational attraction. This will usually produce a super-giant or hyper-giant star which runs short of 'fuel' in a few million years and then undergoes gravitational collapse, produces a

supernova or hypernova and spends the rest of its existence as a black hole.

Accretion disks and gas jets

Most accretion disks and gas jets are not clear proof that a stellar-mass black hole is present, because other massive, ultra-dense objects such as neutron stars and white dwarfs cause accretion disks and gas jets to form and to behave in the same ways as those round black holes. But they can often help by telling astronomers where it might be worth looking for a black hole. On the other hand, extremely large accretion disks and gas jets may be good evidence for the presence of supermassive black holes, because as far as we know any mass large enough to power these phenomena must be a black hole.

Quasars

Visible quasars show a very high red shift. The scientific consensus is that this is an effect of the expansion of the universe between quasars and the Earth. When combined with Hubble's law, the implication is that the quasars are very distant. To be observable at that distance, the energy output of quasars must dwarf that of almost every known astrophysical phenomenon in a galaxy, excepting comparatively short-lived events like supernovae and gamma-ray bursts. Quasars may readily release energy in levels equal to the output of hundreds of average galaxies combined. The output of light is equivalent to one trillion suns.

In optical telescopes, most quasars look like single points of light (i.e. point source) although some are seen to be the centres of active galaxies. Most quasars are too faint to be seen with small telescopes, but 3C 273, with an average apparent magnitude of 12.9, is an exception. At a distance of 2.44 billion light years, it is one of the most distant objects directly observable with amateur equipment. Some quasars display rapid changes in luminosity, which implies that they are small (an object cannot change faster than the time it takes light to travel from one end to the other; but see quasar J1819+3845 for another explanation). The highest red shift currently known for a quasar is 6.4.

Quasars are believed to be powered by accretion of material into super massive black holes in the nuclei of distant galaxies, making these luminous versions of the general class of objects known as active galaxies. No other currently known mechanism appears able to explain the vast energy output and rapid variability. Knowledge of quasars is advancing rapidly. As recently as the 1980s, there was no clear consensus as to their origin. Recent evidence shows that quasars are part of the natural evolution of a galaxy. When the galaxy first form by gravitational collapse of matter into stars the central region of the galaxy was stellar dense resulting in stellar collisions and the formation of star eating black holes. During this early period these black holes themselves coalesced into a single large black hole and with so much stellar material to digest they were very active and gradually grew into super massive black holes. With time as the stellar material at the centre of the galaxy became less dense the black hole only had a periodic lunch. This is what we see today old galaxies close to us have super massive black holes and are relatively inactive, whereas distant (and therefore young) galaxies (quasars) were very active.

Some Important Conclusions

This suggests that the power houses for the mass-energy cycle of the universe were billions of years ago very active in putting mass-energy into the space-vacuum. During this very early period of the universe when galaxies were forming, the energy density of the space vacuum would have been much higher, therefore the high red shift of light from these past distant galaxies (quasars) may not simply because of expansion of the universe but because of decreasing space vacuum energy density from the past to the present. If the space vacuum energy density was higher in the past then all light travelling to us today *will* be red shifted – this is the same effect remember as light leaving a gravitational field (Einstein's gravitational red shift which is the same effect as refraction – the speeding up of light and its increasing wavelength as it moves from a more dense medium to a less dense medium)

In fact the more distant the galaxy, the older it is, and therefore the more the light from that galaxy will be red shifted. Red shift may not be proof that the universe is expanding but that in the past the energy density of the space vacuum was higher. If the red shift of light from more distant galaxies is due to decreasing energy density with time then these distant galaxies are not receding away from us where recessional velocity is proportional to red shift, but are in fact static, in which case the universe is not expanding at all. This does not rule out expansion in the early universe, I believe that the universe did expand during its formation and that the space vacuum energy with time decreased because of expansion and because of universal less energetic energy cycle activity – active quasars evolved into 'dormant' galaxies such as our own the Milky Way.

Dark Energy and an Accelerating Expansion

Recent evidence by two independent astronomical surveys suggests that the universe is expanding at an increasing rate – expansion is accelerating is the current view on the universe. This should not be the case according to conventional theory, the expansion rate of the universe should decrease as the gravity of all the mass in the universe puts on the brakes and slows it down. There are many new theories to explain this, including the invention of what's called dark energy, no one can explain what dark energy is but that it may be responsible for the acceleration by some kind of mysterious new force which is repulsive between matter. Another idea to explain the increase in the rate of expansion is that the effect of gravity over large distances is somehow reduced – it doesn't follow just an inverse square law. This idea can be ruled out because we would notice this in the orbits of planets around our own sun.

The accelerating expansion can be explained in our model of *gravity as an influx of the space vacuum energy density*. If we assume that the universe is still expanding and that the red shift we see is due to both decreasing energy density with time and expansion, then the decrease in the energy density of the space vacuum would mean that the strength of gravity between masses is also decreasing. In either case, expansion or currently static, because of decreasing energy density with time the strength of gravity in the universe is less today than it was billions of years ago. Another interesting thought is that if light from the past, from these distant galaxies, is red shifted because of decreasing energy density, then the light is also speeding up as it travels from the past to the present and since light governs time, in the past time ran slower!

The universe could be much old than we think, relatively speaking of course.

CHAPTER 12
Gravity's Universal Flux

The gravitational curvature of space-time, as a flux of the space vacuum energy density into a mass - where flux rate and density decrease with distance.

One question remained unanswered in my mind for some time, if matter particles are standing waves of electromagnetic energy, why do they end up as a spinning vortices with a twisted configuration when formed in energy mass conversions from free electromagnetic waves. When high energy gamma rays meet and produce matter particle pairs, there must be a some kind of *looping mechanism* for light waves to form matter waves. There is one possible candidate for this mechanism, a little known effect in physics called: The Faraday Effect.

Light's Looping Mechanism in Matter Production

In physics the Faraday Effect or Faraday rotation is a magneto-optical phenomenon, or an interaction between light and a magnetic field. The rotation of the plane of polarization is proportional to the intensity of the component of the magnetic field in the direction of the beam of light. The Faraday Effect was discovered by Michael Faraday in 1845, and was the first experimental evidence that light and magnetism were related. The theoretical basis for that relation, now called electromagnetic radiation, was developed by James Clerk Maxwell in the 1860s and 1870s. This effect occurs in most optically transparent dielectric materials (including liquids) when they are subject to strong magnetic fields.

As the electromagnetic wave travels through a magnetic field the wave rotates by a small angle i.e. its plane of polarization is rotated. The relation between the angle of rotation of the polarization and the magnetic field in a diamagnetic material is:

$$\beta = VBd$$

Where:

β is the angle of rotation (in radians)

B is the magnetic flux density in the direction of propagation (in teslas)

d is the length of the path (in metres) where the light and magnetic field interact

V is the Verdet constant for the material

This empirical proportionality constant (in units of radians per tesla per metre) varies with wavelength and temperature and is tabulated for various materials. A positive Verdet constant corresponds to L-rotation (anticlockwise) when the direction of propagation is parallel to the magnetic field and to R-rotation (clockwise) when the direction of propagation is anti-parallel. If a ray of light is passed through a material and reflected back through it, the rotation doubles. Having discussed the Faraday Effect with a colleague the thought occurred to me that light waves, high energy gamma waves, can therefore interfere with each other magnetically and cause rotation, a twisting of their plane of polarization. This is probably the looping mechanism I had been looking for: where free waves of light energy meet, magnetically interfere with each other by attraction and repulsion along the beam and end up as looped or twisted standing matter waves. This magnetic twisting and looping of the waves means that the free waves can end up as any type of fundamental particle provided there is sufficient energy to form the particle rest mass – which remember is its magnetic angular momentum of spin.

Particle Rest Mass and
Magnetic Moment of Spin

Since particles are standing waves where the fundamental rest mass, the lowest mass-energy configuration, then it must be that more massive particles are oscillating at a higher frequency and therefore must have a greater magnetic moment of spin. Each particle in nature has its own fundamental frequency; this is easy enough to understand in the electron which is a fundamental particle, but what of other particles like protons and neutrons, which are not fundamental because they are made up of quarks? In physics we say that the quark is a fundamental particle, but quarks cannot exist in isolation, they are found inside other particles:

Proton: 3 quarks: 2 up quarks (+) and 1 down quark (-)

Neutron: 3 quarks: 1 up quark (+) and 2 down quarks (-)

Meson: 2 quarks: 1 quark (+ or -) and 1 anti-quark (- or +)

Question: even in high energy collisions between particles quarks do not appear as separate particles, even though we say they are fundamental, so what are they?

Answer: Quarks are not particles at all, but standing wave charge mass centres within a particle.

Since a standing wave proton and anti-proton can be produced from two high energy gamma waves, then each must be made up of a certain standing wave configuration where the oscillating wave is twisted and looped to produce charge-mass centres. We have already seen how an electron standing wave by its twisting and spinning hides its positive field inside the electron giving it a negative field outwards, the same is probably true of the proton except the proton's standing wave configuration probably has three oscillating loops to the standing wave twisting in such a way that they appear as separate charge-mass centres.

The orientation of the field in each loop gives it its fractional charge and the magnetic moment of the spin of each electromagnetic loop, propagating in the loop at the speed of light (c), gives rise to the quark's rest mass. The particle-wave configurations are determined by the interaction of their magnetic fields by the Faraday Effect, it is these interacting magnetic forces which twist and pull the waves into their standing wave shape.

When particles meet in high energy particle accelerators the loops (quarks) are pulled apart into other particle wave configurations which we see as separate particles made of two or more electromagnetic loops (quarks). The quark structure is simply a convoluted electromagnetic wave with lobes showing fractional + and − electric field direction ('charge'). This is why they can not be separated! In collisions the wave rearranges itself. Let's look at the properties of these particles:

1. Electron

Rest mass: $m_0 = 9.11 \times 10{-}31$kg

Frequency of oscillation (f): $f = E/h = mc^2/h$

$$= 9.1094 \times 10^{-31}\text{kg} \times (2.998 \times 108\text{m/s})^2/6.626 \times 10^{-34}\text{Js}$$
$$f = 1.2357 \times 10^{20}\text{Hz}$$

Half Wavelength: $(\lambda/2) = c/2f$
$$= (2.998 \times 10^8\text{m/s})/(2 \times 1.2357 \times 10^{20}\text{Hz})$$
$$\lambda/2 = 1.2131 \times 10^{-12}\text{m}$$

Radius of spin of electron: $r =$ standing wave circumference $(\lambda/2)/2\pi$
$$= 1.2131 \times 10^{-12}\text{m}/2\pi$$
$$\mathbf{r = 1.931 \times 10^{-13}\text{m}}$$

2. Proton

Rest mass: $1.6726 \times 10^{-27}\text{kg}$

Frequency of oscillation (f): $2.2688 \times 10^{23}\text{Hz}$

Half Wavelength: $(\lambda/2)$: $6.607 \times 10^{-16}\text{m}$

Radius of spin of proton: $1.0515 \times 10^{-16}\text{m}$

3. Quarks

Rest mass (up and down): 360Mev

$1\text{ev} = 1.6022 \times 10^{-19}\text{J}$ therefore 360MeV
$$E = 1.6022 \times 10^{-19} \times 360 \times 10^6\text{J}$$

Rest Mass-energy $= 5.768 \times 10^{-11}\text{J}$

By mass-energy equivalence the quark's mass $= E/c^2$
$$= 5.768 \times 10^{-11}\text{J}/(2.998 \times 108\text{m/s})^2$$
$$m_{q0} = 6.4174 \times 10^{-28}\text{kg}$$

Frequency of oscillation (f):
$$= 6.4174 \times 10^{-28}\text{kg} \times (2.998 \times 108\text{m/s})^2/6.626 \times 10^{-34}\text{Js}$$
$$f = 8.705 \times 10^{22}\text{Hz}$$

Half Wavelength: $(\lambda/2)$:

$$= (2.998 \times 10^8 \text{m/s})/(2 \times 8.705 \times 10^{22}\text{Hz})$$

$$\lambda/2 = 1.722 \times 10^{-15}\text{m}$$

Radius of spin: r = standing wave circumference $(\lambda/2)/2\pi$

$$= 1.722 \times 10^{-15}\text{m}/2\pi$$

$$\mathbf{r = 2.741 \times 10^{-16}\text{m}}$$

Summary

Rest Mass	Ring Radius
Electron	
9.11 x 10-31kg	**1.931 x 10^{-13}m**
Proton	
1.6726 x 10^{-27}kg	**1.0515 x 10^{-16}m**
Quark	
6.4174 x 10^{-28}kg	**2.741 x 10^{-16}m**

From the above calculations it can be seen that the higher the rest mass the higher the oscillating frequency of the standing wave particle, this means the fundamental oscillation half wave is shorter and hence the particle standing wave is smaller. (Remember the single proton nucleus which is more massive than the electron is 10,000 times smaller than the standing wave electron oscillating around it)

The quark radius as a separate mass-energy wave comes out 2.6 times bigger than the proton. How can the small proton be made up of three bigger quark standing waves? It makes sense to view the proton as a single standing wave at its own mass energy frequency where by a twisting of the wave through the Faraday Effect it ends up in three mass-energy-charge loops – which we see as quarks.

This explains why quarks can never be isolated as separate particles and are therefore not true fundamental particles – it is quite probable that there are only three true fundamental particles – the neutrino, electron and proton!

These are stable configurations. The rest of the particles which make up our universe are all more energetic convolutions of these – all standing waves of electromagnetic radiation spinning in various ways to demonstrate their properties of charge, mass and spin and it is this spin which we see as inertial mass.

BIG G Verses ZPE

Investigating the Lamb Shift of Electrons by the Energy of the Space Vacuum and Newton's Universal Gravitational Constant, G.

Evers since discovering the existence of Newton's gravitational constant (G) I have been intrigued and puzzled by what this universal constant actually means.

Used in Newton's formula:

$$F = - GMm/r^2$$

It determines the magnitude of the gravitational field force exerted on a mass (m) at a distance (r) from the centre of a gravitating mass (M). And since the gravitational acceleration (g) of any mass independent of type of substance in this field is given by:

$$g = GM/r^2$$

This universal constant may be fundamental to understanding what gravity really is. What does this constant G mean? What property of the universe gives it its value?

$$G = 6.67390 \times 10^{-11} Nm^2 kg^{-2}$$

Since we multiply the Mass (M) by G this tells us that the gravitation force/acceleration is proportional to G. The bigger the value of G the greater the gravitational acceleration – the gravity field becomes stronger. Let us look at the units for G, maybe there is a clue there:

Rearranging the first equation in terms of G gives:

$$G = Fr^2/Mm$$

Reducing the right side of the equation to units gives:

$$G = Nm^2/kg^2$$

G has units of Newton metre squared per kilogram squared. What does this mean? This does not help much, but there is some hope in the units m^2, which is equal to area (A):

$$G = \text{force} \times \text{area} / \text{mass squared}$$

Let's try reducing the equation to fundamental base units:

Since by $F = ma$, the Newton unit is equal to: kgm/s^2

Substituting these units into the equation we get:

$$G = kgm/s^2 \times m^2 / kg^2 = m^3/kgs^2$$

This is interesting since m^3 is volume! We now get:

$$G = volume / mass \times time\ squared$$

This is also very interesting having the units of area and volume in the units for G since these dimensions have significant meaning in the gravitational energy flux model. Let's look at the second equation:

$$G = gr^2/M$$

Remembering that g has units of acceleration (m/s^2) gives:

$$G = m/s^2 \times m^2 /kg = m^3/kgs^2$$

Here G has units of metre cubed per kilogram second squared - the same as in the first equation. Does *this* help us understand G? There area and volume (V) are tangible physical dimensions of space-time and space-vacuum energy density flux. Another interesting observation is that in the equation Volume/Mass appears, which is the inverse of density:

$$Density = Mass/ Volume$$

Therefore the dimensions of G become:

$$G = 1/density \times time\ squared$$

This equation is intriguing since it contains the term density and links to our gravitational model which relies on the accelerating flux of the space vacuum energy density ρ_v. The term per time squared is also interesting as this suggests an increasing change with time – acceleration.

Special Note on G

Once this book had been written, the meaning of the gravitational constant (G) still bothered me. The above link to mass, volume, density and time squared seem to be satisfying to a point, but not completely. On return to the problem I discovered something more interesting about G and something which may be another cornerstone for the gravity mechanism model:

Here is another derivation for the units of G:

$$F = G\,Mm/r^2$$

Since: $F = G\,Mm/r^2$

Therefore: $G = Fr^2/Mm$

In Base Units:

$$F = kg.m/s^2$$

Therefore: $G = kg.m/s^2.m^2.1/kg^2$

Simplifying gives:

$$G = m^3/s^2 \text{ per kg}$$

G = Volume/Time Squared per mass

In Physical Quantities

G = <u>Rate of Change of Volume</u> per unit mass

This equation shows that the constant 'G' represents a rate of change of volume per unit mass. This fits in perfectly with the model of a collapsing sphere of the space vacuum energy into a mass. It seems that:

G is a constant which determines the rate of change of contracting space volume -for a given amount of gravitating mass.

Planck's constant (h) and the Fundamental Nature of the Universe

Planck's constant h is a small energy unit (quantum): If the frequency of oscillation of an electromagnetic wave equals one, one oscillation per second, then by:

$$E = hf,$$

The energy of this low frequency photon

$$= 6.626 \times 10^{-34} J.$$

Planck's constant has the units Js which are the units of angular momentum. Like the other fundamental constants such as the speed of light (c) and the Gravitational constant (G), Planck's constant is a fundamental or natural unit based on the very physical properties of the universe. In other words it is the physical structure of the universe which determines the values for these constants and therefore conversely these constants must also describe the physical nature of the universe. If we assign values of unity (one) to the fundamental base units of mass, length and time, charge and temperature then other fundamental base (Planck) units can be derived:

(1) Planck Length Lp:

$$Lp = SQR(hG/2\pi c^3)$$

$$Lp = SQR[6.62606876 \times 10^{-34}Js \times 6.67390 \times 10^{-11}Nm^2kg^{-2}/2 \times \pi \times (2.99792458 \times 10^8 m/s)^3]$$

Planck Length = 1.61624 x 10⁻³⁵m

Planck length is *theoretically* the smallest possible unit of length and therefore represents *theoretically* the tiniest finite structure of space-time. It is also *theoretically* the radius of the smallest particle which can exist – a Planck Particle.

(2) Planck Mass Mp:

$$Mp = SQR(hc/2\pi G)$$

Planck Mass = 2.17645 x 10⁻⁸kg

This represents the mass of the most massive and therefore smallest particle which can *theoretically* exist: a Planck Particle.

(3) Planck Time (Tp):

$$Tp = SQR(hG/2\pi c^5)$$

Planck Time = 5.39121 x 10⁻⁴⁴s

All three, Planck length, mass and time resent the fine structure of space time. It is *theorised* that Planck particles makeup the fabric of space-time. It is very interesting to note that the speed of light (c) through a space vacuum matrix of Planck particles can be derived from Planck units:

Since velocity = distance /time

$$c = Lp/Tp$$

$$= 1.61624 \times 10^{-35}m/5.39121 \times 10^{-44}s$$

Speed of light in space vacuum

$$= 2.9979167 \times 10^8 m/s$$

Now this makes sense if we assume it takes Planck Time (Tp) at the speed of light to traverse the Planck particle. If we consider that the space vacuum is filled with Planck Particles then we can calculate the density of this medium: the Space vacuum mass density:

Since density = mass/volume

The Planck Density ρ_p:

$$\rho_p = Mp/Lp^3$$
$$= 2.17645 \times 10^{-8} kg/(1.61624 \times 10^{-35}m)^3$$

$\rho_p = 5.155 \times 10^{96} kg/m^3$

This therefore gives the density (10^{96}) of the space-vacuum matrix, and as said before very much greater than nuclear densities.

The Energy Density (energy per volume) of empty space must therefore by $E = mc^2$ equal:

$$\rho v = MpC^2/Lp^3$$

Energy density is the mass density times c^2

$\rho v = 1.54543 \times 10^{113} J/m^3$

This is a massive theoretical energy density for space-time, based on the natural units of the physical universe. If we compare the electron to the Planck particle:

Electron	Planck Particle
Length:	
$10^{-12}m$	$10^{-35}m$
Mass:	
$10^{-31}kg$	$10^{-8}kg$

We see that the Planck particle is much smaller than the electron and much more massive, this means that the theoretical Planck particle is much denser than ordinary matter. If Planck units really define the smallest quantum structure of the universe, then the fundamental particulate quantum structure of space-time is super dense and super small compared to ordinary matter particles.

Matter is made of fundamental mass particles, these particles have a much bigger wavelength and size to the Planck particles of the fabric of space-time and therefore like light through glass are transparent to the medium, which is why to us mere matter mortals space appears empty. We can easily wave our hand through the vacuum of space even though it has an enormous energy density.

By Newton's second law the space vacuum exerts a drag force on <u>accelerating</u> objects through this medium and likewise if the medium of the space vacuum <u>accelerates</u> through an object it will exert a drag force on the object which we see as gravitational force.

By rearranging the above equations we can derive G from Planck's fundamental units of mass, time and length, but the units of Planck measurement were derived using G and so this circular derivation for G has no real value in our understanding of G. Using the equation for Planck length:

$$G = 2\pi Lp^2 c^3/h$$

But can we derive it from some meaningful physical observation and measurement of the universe? In our gravity model an accelerating flux of space-vacuum energy into a mass produces all the known effects of a gravitational field, therefore somewhere in this model there must be a way to obtain directly the value of G. Is there a link between G, gravitational space vacuum energy density flux and the spinning vortex nature of matter particle standing waves which cause this flux? If indeed fundamental particle standing waves spinning and convoluting at the speed of light are the true cause for space-time flux into matter, then G must have some significance in the model. (This we have seen above in the base units for G where they are equal to a contracting and accelerating sphere of space vacuum volume.)

G is the constant which determines the strength of a gravitational field – the acceleration of the space vacuum energy into matter particles. In exploring for an answer to this question I recently returned to the electron vortex model puzzling over the Lamb Shift of electrons oscillating about a nucleus in an atom. If you recall Lamb Shift is the energy/frequency increase ($\Delta E/\Delta f$) in electron orbital energy levels, due to the presence of virtual electron and positron pairs and annihilation photons which make up energy quantum fluctuations of the space vacuum – the electrons jiggle in their orbit due to the presence of these short lived particles.

Lamb shift is evidence for the space vacuum energy density. The Lamb shift frequency change (Δf) in the hydrogen ground state electron has recently been measured with greater accuracy and was found to be of the value:

8172.86 Mhz for the hydrogen atom

8184.00 MHz in the deuterium atom.

The electron whizzing around the nucleus encounters and absorbs the energy of these virtual and real energy particles and each time it does so its own energy and thus frequency jumps from the lowest energy state (ground state) up to an unstable higher energy level – this we see as the Lamb Shift. Almost immediately the electron jumps back down and releases the energy as a quantum of electromagnetic wave energy – a photon. The quantum of energy absorbed equals the quantum of energy emitted. There is a continuous exchange of energy between electrons and the space vacuum. The previously calculated wavelength of a photon which represents the energy jump of Lamb Shift:

Wavelength of particle or photon = 3.671cm

The temperature of this radiation we calculated as: T = 0.0789K. This is below the temperature of the cosmic background radiation of 2.725 K, but is within the CMBR spectrum.

Quantum Energy of the virtual particles or emitted photons

$$= 5.4154 \times 10^{-24}J$$

Mass of the absorbed virtual particles

$$= 6.0171 \times 10^{-41}kg$$

The virtual mass-energy of the observed Lamb Shift is extremely low compared to the mass of a real electron: $9.11 \times 10^{-31}kg$ and because these energy quanta have very low mass they have a large size:

Wavelength size of virtual quantum = 0.03671m

Wavelength size of real electron = $2.426 \times 10^{-12}m$

The virtual particle is 10^{10} times bigger than the electron, therefore since these virtual particles of the space vacuum have little mass-energy and have a much bigger wavelength than real matter particles, then matter particles are transparent to this medium. The picture gained from all of this is that the space-time matrix is made of an almost infinite theoretical source of tiny but massive Planck particles and a sea of virtual particles and real photons – a truly bubbling sea of mass energy.

Existence time of the virtual particles

$$\Delta t = 9.737 \times 10^{-12} \text{seconds}$$

These virtual particles by Heisenberg's uncertainty principle can only exist for a very short time – try and picture the surface of energetic choppy water at the microscopic level – as we zoom in first we see splashes of water droplets leaving the surface, these droplets have sufficient kinetic energy to leave the surface but soon fall back down to the mass of water – they exist for a very short time. If a water droplet had sufficient energy and reach velocities greater than escape velocity it would leave the sea and become independent water 'particle.' This is like a virtual particle absorbing a photon with sufficient energy to make it massive enough to become a real particle. Now let us zoom in closer and we see the individual water molecules evaporating from the surface and like the water droplets they will return to the mass of water if they do not have enough energy to escape.

In this simple analogy of the space vacuum the water droplets are the virtual particles being absorbed by matter (Lamb Shift) and the water molecules are the Planck particles of the space-time matrix.

Returning to the problem of whether G can be derived from the mechanics of this gravity model, which at its heart is the interaction of electrons with the space vacuum, then the Lamb Shift of electrons seems the best place to start and therefore I decided to calculate the ratio of how much energy is absorbed per quantum shift of the hydrogen electron compared to the electron's own mass-energy:

Ratio of absorbed gravitational Lamb Shift energy (ΔE_L) to mass energy of the absorbing mass particle – the hydrogen electron (E_e):

$$G_L = \Delta E_L / E_e$$

Let us calculate this as accurately as possible for the simplest fundamental particle: a hydrogen atom which has one electron in the ground state being Lamb shifted by 8172.86 Mhz:

$$\Delta E_L = hf_L$$
$$= 6.62606896 \times 10^{-34} \text{Js} \times 8172.86 \times 10^6 \text{s}^{-1}$$

$$\mathbf{\Delta E_L = 5.415393396 \times 10\text{-}^{24} J}$$

The electron has a mass energy E_e of:

$$E_e = mc^2$$

$$= 9.10938188 \times 10^{-31}kg \times (2.99792458 \times 10^8 m/s)^2$$

$$\mathbf{E_e = 8.187104 \times 10^{-14}J}$$

Therefore:

$$\mathbf{G_L = \Delta E_L / E_e}$$

$$= 5.415393396 \times 10^{-24}J / 8.187104 \times 10^{-14}J$$

$$\underline{\mathbf{G_L = 6.61 \times 10^{-11}}}$$

The magnitude of the ratio of the Lamb Shift mass-energy absorbed by an electron to the mass-energy of the electron comes out very close to the magnitude of the universal gravitational constant G (6.67 x 10^{-11}). This could be coincidental, but it is nevertheless very interesting to think that the magnitude of the gravitational constant G is a ratio of the Lamb Shift energy absorbed by an electron to its ground state mass-energy.

<u>This would suggest that the interaction of electrons with the space vacuum determines the magnitude of G!</u>

Noticing that the magnitude for the derived G_L comes out slightly less than the measured value for G, I reasoned that this was because I calculated G_L for the simplest and least massive particle: the hydrogen atom, thus I tried the calculation again using the Lamb Shift frequency of heavy hydrogen (Deuterium) = 8184.00Mhz. This atom has an extra neutron in the nucleus compared to normal hydrogen:

For Deuterium:

$$\Delta E_L = hf_L$$

$$= 6.62606896 \times 10^{-34}Js \times 8184.00 \times 10^6 s^{-1}$$

$$\Delta E_L = 5.422774837 \times 10^{-24}J$$

Therefore:

$$G_L = \Delta E_L / E_e$$

$$= 5.422774837 \times 10^{-24}J / 8.187104 \times 10^{-14}J$$

$$\textbf{For deuterium } G_L = 6.62 \times 10^{-11}$$

This comes out higher for deuterium than hydrogen and closer to the exact magnitude of G:

$$\text{Magnitude of } G = 6.67390 \times 10^{-11}$$

Since the Lamb shift frequency of an electron is dependent on the type of atom and probably the electron configuration of that atom then the ratio G_L is also dependent on the type of atom. It follows therefore that if G_L is equal to G then G_L must be an average value for all the electron Lamb shifts in a gravitating mass:

$$\mathbf{G_{Lav} = G}$$

This makes sense if we consider than in matter there are many different types and masses of atoms with electrons shifted from the ground state by the energy of the space vacuum – atoms with different proton numbers (z), electron numbers and mass numbers (A) resulting in different electron configurations and therefore slightly different Lamb Shift frequencies caused by the presence of the space vacuum energy.

The magnitude of G may be an average value based on the ratio of Lamb the shift of fundamental particle in matter to their mass-energy.

Question: what is the average energy and frequency of Lamb Shift based on the possible link to G?

Answer: Since: $G_L$average = G = ΔE_Laverage$/E_e$

Then the average Lamb shift of electrons in matter equals

$$\Delta E_L \text{ average} = G \times E_e$$
$$= 6.67390 \times 10^{-11} \times 8.187104 \times 10^{-14}\text{J}$$

$$\mathbf{\Delta E_L \text{ average} = 5.46399 \times 10^{-24}\text{J}}$$

With a frequency of:

$$f_L = E/h$$
$$= 5.46399 \times 10^{-24}\text{J}/6.62606896 \times 10^{-34}\text{Js}$$

$$\mathbf{f_L = 8,246.204MHz}$$

Slightly higher than hydrogen's 8172.86 MHz due to the range of Lamb Shifts in matter. Now *assuming* that G_L is the magnitude of the average Lamb shift for electrons in matter, the gravitational acceleration of a mass equals:

$$g = G_L M/r2$$

Or better still:

$$g = \Delta E_L/E_e \times M/r^2$$

If this relationship is true then it shows that gravitational acceleration is directly proportional the Lamb shift of electrons, or more correctly to the Lamb Shift of matter particles due to energy exchange with the space vacuum. This makes some sense: g depends on how much space vacuum (ZPE) energy is absorbed per unit mass of matter (G_L) and how much matter there is (M). Intrigued by the possible relationship between gravity and the energetic space vacuum I carried out further research on the Lamb shift and came across this equation which calculates the Lamb shift magnitude directly from fundamental constants, the mass of the electron m_e and principal quantum numbers :

$$\Delta E_L = \alpha^5 m_e c^2 \times [k(n,0)/4n^3] \text{ for } L_{qn} = 0$$

Where: α = the fine structure constant

The fine structure constant is a dimensionless number and a constant which determines the <u>strength of electromagnetic interactions.</u> Since the electron is Lamb shifted by its electromagnetic interaction with the space vacuum it makes sense that it appears in this equation. It can be derived from electron charge e, the speed of light c, the electric permittivity of free space ε_0 and Planck's constant h:

$$\alpha = e^2/2\varepsilon_0 hc = 7.2974 \times 10^{-3}$$

It is also interesting to note that the electrical *permittivity of free space*, a property of the space vacuum which determines the *strength* of electric fields is also present in the formula.

The Lamb Shift energy is proportional to the electromagnetic interaction constant

In this formula, m_e = mass of the electron: We have already seen how this physical quantity is important, the electron's rest mass is important in determining how much its mass-energy is shifted by the space vacuum energy

$$m_e = 9.10938215 \times 10^{-31} kg$$

$$c = \text{speed of light} = 2.99792458 \times 10^8 m/s$$

In the formula, k (n, 0) is a numerical factor which varies slightly with n from 12.7 to 13.2

n = is the principal quantum number (n = 1, 2, 3 ...)

This is the Bohr model of the electron orbital energy levels, the orbital shell number of the electrons: for the first shell where the electron in a hydrogen atom rests in the ground state n = 1.

L_{qn} = is the azimuthal quantum number

This the *square* of the *orbital angular momentum* (L^2) and takes on values of: 0, 1, 2, ... L_{qn} is equal to n-1. So when n = 1 then L_{qn} = 0 as in the above case where the Lamb shift formula is used.

Interesting that an equation that calculates the magnitude of Lamb shift should have a constant relating to electron orbital angular momentum!

Let us now use this formula to calculate the Lamb Shift for the hydrogen atom (Electron shell 1) where n =1 and L_{qn} = 0: First of all we will take an average value for k (n, 0) equal to 12.95 and assign a value to the multiplying constant $[k(n,0)/4n^3]$ in the equation and call this k_L:

$$k_L = [k(n,0)/4n^3]$$

$$= 12.95/4$$

$$\mathbf{k_L = 3.2375}$$

Therefore:

$$\Delta E_L = \alpha^5 m_e c^2 \times k_L$$

$$= (7.2974 \times 10^{-3})^5 \times 9.10938215 \times 10^{-31} kg \times (2.99792458 \times 10^8 m/s)^2 \times 3.2375$$

$$\mathbf{\Delta E_L = 5.485054 \times 10^{-24} J}$$

This comes every close to the latest measured value of the Lamb shift: $\Delta E_L = 5.415393396 \times 10^{-24} J$. So this formula actually works. If we calculate the value in this case of k(n, 0) using the measured value for Lamb Shift it comes to 12.7854 giving k_L equal to 3.19635. Now knowing that with this formula we can calculate Lamb shift, what value does it give for G?

Let us look at the Lamb shift formula again:

$$\Delta E_L = \alpha^5 m_e c^2 \times k_L$$

Rearranging this formula gives:

$$\mathbf{\Delta E_L/m_e = \alpha^5 c^2 \times k_L}$$

Arranged in the terms of Lamb shift energy over electron mass is very similar to our formula for calculating G:

$$G = \Delta E_L / E_e$$

If we substitute the electron mass m_e for electron mass-energy Ee using Einstein's $E = mc^2$ we get:

$$\Delta E_L c^2 / E_e = \alpha^5 c^2 \times k_L$$

The c^2 cancels out on both sides giving:

$$\mathbf{\Delta E_L / E_e = \alpha^5 \times k_L}$$

This seems to be exactly what we want, but does it give us a value for G as predicted before using $G = \Delta E / Ee = 6.67390 \times 10^{-11}$?

Let us see:

$$\Delta E_L / E_e = \alpha^5 \times k_L$$

$$= (7.2974 \times 10\text{-}3)^5 \times 3.19635$$

$$\mathbf{\Delta E_L / E_e = 6.615 \times 10^{-11}}$$

The Lamb Shift equation based on the fine structure constant of electromagnetic interactions gives a value slightly less than G as before when we calculated GL from the Lamb shift data.

If k_L were known as an average for all the Lamb Shifts present in matter, then the equation may give us an exact value for G and if Lamb Shift and the fine structure constant are indeed a fundamental part of the mechanism for gravity, then the following equation may be the solution to calculating gravitational field strengths from the fundamental quantum mechanical constant: the fine structure constant.

$$\mathbf{G_L = \Delta E_L / E_e = \alpha^5 \times k_L}$$

It is now not surprising that our ratio of lamb shift energy to electron mass energy is a dimensionless number since both the fine structure constant and the quantum constant K_L are also dimensionless. GL is a ratio since it does not matter whether you examine the ratio of lamb shift energy to particle energy or lamb shift mass equivalent to particle mass – both are ratios which give you the magnitude equivalent to G.

In this equation we have the magnitude of the universal gravitational constant linked to lamb shift, electron particle mass-energy, the fine structure constant which determines the strength of electromagnetic interactions (between the space vacuum and fundamental particles like electrons) and orbital quantum mechanics.

We can re-write the equation for gravity as:

$$F = -G_L Mm/r^2$$

Or

$$F = -\Delta E_L/E_p Mm/r^2$$

Or

$$\mathbf{F = -\alpha^5 k_L Mm/r^2}$$

In each case we are assuming that we are using the average value for all the space vacuum energy Lamb shifts in the fundamental matter particles (p), electrons and quarks, of a gravitating mass.

What is fundamental to note here, is that if the above equation holds true, then this equation links Einstein's gravitational field theory to quantum mechanics – <u>a unification of physics</u>.

Summary:

1. The space time matrix of Planck particles is filled with an enormous amount of short lived virtual particles which cause the Lamb Shift of electrons in orbit around an atom.

2. The mass energy density of Planck space and the mass energy density of the space vacuum virtual particles limit the magnitude of the speed of light to c.

3. Matter is continually absorbing and emitting the energy of these virtual particles from the space vacuum.

4. Matter particles momentarily take on an excited state – become Lamb Shifted

5. These excited particles continually emit the absorbed space vacuum energy back to the space vacuum as microwave radiation and into matter as heat.

6. There is a cyclic flux of energy from the space vacuum into matter and back again in radiative and non-radiative forms.

7. It is this cyclic flux of space vacuum energy into matter which we see as gravity.

8. The magnitude of the absorbed mass energy per mass energy of the electron ($G_L = \Delta E_L / E_e$) dictates the magnitude of G and hence the strength of the gravitational field around a space vacuum interacting mass.

9. The fine structure constant which determines the strength of electromagnetic interactions may be related to the strength of gravitational fields through the formula:

Quantum Gravitational Field Theory Formula

$$F = -\alpha^5 k_L Mm/r^2$$

10. If this equation proves correct, then it unifies field theory and quantum mechanics.

CHAPTER 13
Time to Travel

The quest for the stars is afoot.
Space-time mechanics and our understanding of gravity
is key to travelling through space and time.

Space-time travel across the universe to distant stars and galaxies is only possible if we can harness the true forces of nature which govern the fabric of space, time and gravity itself. Travelling to the stars at breakneck speeds, faster than the speed of light (superluminal) may be possible, but only if we understand what truly governs the speed of light. We are no longer in the battle to fly machines through empty space but through a universal space-vacuum that limits mortal velocity and the speed of light. In return this universal speed barrier, dictated by the space vacuum energy density also governs time. Is it possible therefore to harness and manipulate the very fabric of space-time and travel to the stars?

Without matter and the space vacuum mass-energy there is no space-time. The space vacuum energy density limits the speed of light and governs time.

The Gravity Warp

We have seen that in this model for gravity that empty space is not some mystical dimensional mathematical fabric of the universe but a tangible and very real space-vacuum mass-energy density medium – the energy which keeps the universe inflated under vacuum pressure and in existence. This source of energy which is in a constant cycle between matter and space through the massive power houses of the universe is what keeps matter itself in existence – the mass-energy cycle sustains the universe. This flux of space vacuum energy into matter is what we see as gravity, is a very real source of energy: gravitational energy. This energy density flux into all matter is part of the universe's energy cycle, keeps everything in order and drives the nuclear processes in stars which are under gravitational collapse and the stars and planets together on orbit about each other. They key question is can we harness this energy flux or even manipulate it?

Imagine a stone falling to the ground taken by the flux of the space-vacuum, what if we could reverse this flux or interfere with it? Would the stone's acceleration of free fall change? One imagines to do this would be seen as anti-gravity. Okay, let's take the stone out into 'empty space' and by some means induce the space-vacuum energy density to flux, would the stone move with the flux? If this model of gravity is correct, then yes, it would. To move a particle through space that's all we have to do is change the energy density around the particle, and cause a flux of energy through the particle in the direction we want to move.

Question: So how do we manipulate the energy density?

Answer: The same way matter does.

Modern physics suggests that to design a spaceship to travel up to near light speeds we have to warp space in front and behind the ship – in essence make a gravitational wave for the space ship to surf upon. The spaceship itself is lying in flat space-time where there is no acceleration, inertial or space-time dilation affects on the crew or spaceship. The spaceship simply rides a wave of warped space-time.

Conventional Physics

Configurations of space-time curvature for a warp drive

Flat Space-time or 'Empty' Space-Time: with no gravitational field, this equals zero 'g' where there is no space time distortion.

Compression Well of Space-Time: In front of the ship we need to create a region of distorted space-time of a normal negative gravitational field, negative 'g') – the same as the gravity field around a mass. In this region space and time are distorted so that the spaceship falls into this well or trough – the ship is pulled towards this region - accelerates towards it.

Rarefaction Hill of Stretched Space-Time (Trough): Behind the ship we need a region of distorted space-time of a positive gravitational field, positive 'g'. Here our space ship accelerates away from this region – it is pushed.

1. Flat Space-Time: Zero 'g'

2. Compression Well: Negative 'g'. Wave Trough, attraction.

3. Rarefaction Hill: Stretched Space-Time. Positive 'g', repulsion. Wave Crest

If this were possible the spaceship would be continuously attracted, accelerate, to the negatively distorted space in front of the ship and repelled, accelerate away from, the positively distorted space behind. The acceleration of the spaceship would be directly proportional to the amount of space-time distortion. The beauty of this type of space-ship is that it could travel any where in the universe – even in empty space.

Creating an Artificial Gravitational Field

Matter distorts space-time and creates a negative gravitational field where other small masses placed in this field are seen to accelerate towards the larger mass. The problem is it requires an awful lot of matter to create even a weak gravitational field: the whole mass of the Earth at its surface only generates a g-field of $9.81 m/s^2$. Even if we could generate a negative g-field this strong and our spaceship was accelerated by 9.81m/s every second to get to near light speed would take:

$$1 \text{ Year} = 60 \times 60 \times 24 \times 365.24s = 3.1557 \times 10^7 s$$

$$\text{Since } a = (v-u)/t$$

Initial speed u is zero therefore:

$$t = v/a$$

$$\text{Since } v = c \text{ and } a = g \text{ then:}$$

$$t = c/g$$

$$= 2.998 \times 10^8 / 9.81$$

Time to reach light speed at $9.81 \ m/s^2$

$$= 3.056 \times 10^7 s$$

$$= \text{almost one year!}$$

During this time to accelerate to light speed our space ship would have travelled:

$$\text{Distance} = \frac{1}{2} at^2$$

$$= \frac{1}{2} \times 9.81 \times (3.056 \times 107)$$

$$x = 4.581 \times 10^{15}m$$

Distance travelled to reach light speed $= 4.6 \times 10^{12}km$

The distance travelled at the speed of light (1 Light Year):

$$d = vt$$

$$d = 2.998 \times 10^8 m/s \times 3.1557 \times 10^7 s$$

$$1 \text{ Light Year} = 9.461 \times 10^{12}km$$

In other words accelerating to the speed of light at g we would have travelled a distance of ½ a light year. If we wanted to reach the nearest star Proxima Centauri which is approximately 3.5 lights years away we would accelerate at g for 1 year, cruise at near light speed for 2.5 years and then decelerate at g for another year, giving us a total travel time of 4.5 years. This is a feasible travelling time. With maybe one year on exploration and renewing of resource from any orbiting planets around Proxima Centauri this would give us a round trip of ten years – and this is just the nearest star! Burning energy on our space ship for acceleration and deceleration time of 2 years would require an awful amount of fuel. It is clear we need an efficient system to reduce the fuel payload – ion drives driven by nuclear fission (or in the future fusion) at the present is the only answer, but the acceleration rates of ion drives fall far short of even an acceleration rate of g. To make star travel possible we need speeds greater than light speed i.e. superluminal velocities. But is it possible to build a warp drive? Yes, if we can generate gravitational effects on space-time.

Mass warps space-time

This falls along lines of current thinking and here Einstein comes to the rescue because today we don't just talk about mass or energy but mass-energy.

By current reasoning an energy field has mass and therefore gravity. (Not all scientists are convinced of this).

Mass-energy warps space-time

If this is true then we need to create a very powerful energy field to warp space-time because a large amount of energy is equivalent to only a small amount of mass. To create a gravitational field equivalent to that generated by a tiny1 kg mass, we need an energy field of:

$$E = mc^2 = (1\text{kg} \times 3.00 \times 10^8 \text{m/s})^2 = 9 \times 10^{16} \text{Joules!}$$

At a distance of 1m away from this mass-energy field g would be equal to only:

$$g = GM/r^2 = 6.673 \times 10^{-11} \text{ m/s}^2$$

This is a very feeble 'g' acceleration compared to that of the Earth's. To create a g field equivalent to that at the Earth's surface (9.81m/s^2) we need the mass-energy equivalence of:

Mass of Earth $= 5.976 \times 10^{24}\text{kg}$

Therefore energy equivalence:

$$E = mc2$$

$$= 5.976 \times 10^{24}\text{kg} \times (2.998 \times 108\text{m/s})^2$$

Energy field to generate g of: $5.37 \times 10^{41}\text{J!}$

This is a ridiculous amount of energy to contain in a field and is unfeasible. What is also true is that this field energy would generate huge a space-time curvature (we are using the mass size of the Earth remember to generate g at the surface, for our spaceship we do not need to create such as huge energy field because by Newton's Law:

$$g = GM/r^2$$

Since g is proportional to $1/r^2$ if we reduce r then g increases, so by reducing the size of the energy field, to around the size of a spaceship, we can reduce the mass-energy to produce the same g:

Say the field generated outside our spaceship is 100m across then the mass-energy needed to produce a gravitational acceleration of g would be:

$$M = gr^2/G$$

$$= 9.81 \times (100\text{m})^2/6.673 \times 10^{-11}$$

$$M = 1.47 \times 10^{15}\text{kg}$$

Which is equivalent to 1.32×10^{32} Joules

This is a highly concentrated energy field and would still require enormous power to maintain. If this field was electrically driven the power consumption would be 10^{32} Watts! Leaving aside the problem for a moment that vast amounts of mass-energy are needed to warp space-time, let's think about what type of energy field we could use:

The most obvious is an electromagnetic energy field. Electromagnetism comes to the rescue this time for an electromagnetic field stores electromagnetic energy – and according to many scientists this type of field because it has mass-energy generates gravity. A simple battery operated electromagnet stores energy in its magnetic field and therefore the field has mass-energy and gravity.

Electromagnetic Space-Time Gravity Machines

To maintain the electromagnetic field requires an energy supply, the battery. Let us imagine our electromagnet runs at 5 amps of current from a 15 Volt supply; this represents a power consumption of

$$Power = Voltage \times Current = 15V \times 5A = 75 \ Watts$$

This is an energy consumption of 75 Joules per second and would produce a magnetic field strength (B) of:

$$B = \mu nI$$

Where B is the magnetic field strength in Teslars (T), u is the magnetic constant, n is the number of coil turns per metre and I is the current.

A 1000 turn coil of 0.25m long would give n 4000

$$B = 1.2566 \times 10^{-6} \times 4000 \times 5A = 0.0251 \ Tesla$$

Even though our electromagnet has a magnetic field that is many times stronger than the Earths its mass energy equivalence and therefore gravitational effect is extremely small. A magnetic field as high as 40 Tesla, which has yet to be achieved would require a current of almost 8000 amps and a voltage of about 24,000V! Let's imagine this is possible (using superconductors would reduce the power needed) then the energy consumption, disregarding heat loss, would be 192 million Joules per second!

The mass equivalence of this magnetic field by $E = mc^2$ would be
$$2.133 \times 10^{-9} kg$$

One metre away from the electromagnet the g field would be 1.42 x 10^{-19}m/s^2

A very feeble g field indeed compared to the Earths 9.81m/s^2

So it seems even if we build the most powerful electromagnet ever built the space-time distortion in its energy field would be hardly detectable, except by a very sensitive measuring technique called laser interferometry and all this assumes that an energy field can warp space-time. In my model for gravity only mass particles can distort space-time in terms of creating a flux of the energy vacuum.

Travelling at Near Light Speeds

If we manage to reproduce the gravitational effects of mass by an induced flux of the space vacuum energy density travelling to the stars at near light speeds is a real possibility. Our spaceship design using the concept of vacuum energy fluxing will work in much the same way that modern jets work in air, instead of a flux of air through the jet the exhaust of which thrusts the jet forward by Newton's Third Law of Motion, action and reaction, our spaceship will pull space-time vacuum energy through an electromagnetic vortex drive. <u>This principle is similar to water jet skis and smoke rings.</u> Is it possible to accelerate this type of space vacuum energy vortex drive spaceship at velocities greater than the speed of light? Not according to Einstein's theories on light, but there may be a way:

Remember the speed of light is governed by the space-vacuum energy density, if we could slip stream our space ship electromagnetically through this sea of mass-energy particles we could travel faster than light speed!

Travelling Through Time

Changing the space vacuum energy density affects the speed of light and hence time, so again time travel is possible.

A low space-vacuum energy density: the speed of light increases and therefore time runs faster

A high space-vacuum energy density: the speed of light decreases and therefore time runs slower.

It seems therefore that controlling the dynamic flux and energy density of the space-vacuum is key to space and time travel.

CHAPTER 14

Understanding of the Nature of the Universe in the Light of Gravitational Flux Theory

1. Black Hole Singularities

Since the prediction and discovery about black holes one thing has certainly bothered me over the years – the singularity! It seems inconceivable, even with the wildest imagination that the massive core of a giant star could be gravitationally squashed beyond nucleon degeneracy pressure to a mathematical point of infinite density. During the sudden collapse of a star the gravitational forces on the core have to overcome first, electron degeneracy pressure. When this happens the electrons are smashed into the protons forming a neutron star, (a neutron is mad of a proton and electron) if the original star is big enough then the gravitational forces will continue to squeeze the core even smaller, pushing the neutrons together until the final barrier, neutron degeneracy pressure is overcome, beyond this there are no known laws in physics to prevent the core collapsing to a point – a singularity of infinite density.

Let us re-examine this black hole forming process in the light of classical physics and the theories and physics expressed in this book. When the core of a collapsing star reaches the size of what's called the Schwarzschild Radius (R_s). The core reaches a point where the gravitational field around it is so intense that light can no longer escape and therefore it becomes a true black hole. The size of the Schwarzschild radius depends only on the mass (M) of the shrinking core:

$$R_s = 2GM/c^2$$

Where G and c are constants. If the mass of the core is greater than 1.4 solar masses and less than 2.5 it will form a neutron star, a core mass of greater than 2.5 solar masses will form a black hole. Let us examine the collapse of a core which is exactly on the limit of 2.5 solar masses:

The Schwarzschild radius will be:

$$Rs = 2 \times 6.673 \times 10^{-11} \times (2.5 \times 1.99 \times 10^{30}\text{kg})/(2.998 \times 10^{8}\text{m/s})^{2}$$

Rs = 7,387m = 7.4km!

At this point the gravitational field at the surface of the core equals:

$$g = GM/r2$$
$$= 6.673 \times 10^{-11} \times 2.5 \times 1.99 \times 10^{30}/(7.287 \times 10^{3})^{2}$$

g = 3,693.6 m/s^2

The escape velocity from its surface is the speed of light! The star is incredibly dense: 2.946×10^{18} kg/m^3 – 17 times denser than the extreme density of an atomic nucleus of density 1.72×10^{17}kg/m^3! The core has become a giant nucleus of neutrons being squashed together tighter than the nucleon particles in a nucleus and as it shrinks a fraction more smaller than the Schwarzschild radius to become a black hole the core will fail to radiate light. In classical theory not even light can now escape this super dense core it is on the verge of becoming a black hole and if the core is a fraction more massive it will continue to shrink under gravity to a singularity, but is there anything in nature to stop this total collapse? I believe there is and the answer to this question comes from quantum mechanics and the wave-particle duality nature of matter. We must remember that we are not just squashing particles but particles which have a wave nature – standing waves of electromagnetic energy oscillating at the speed of light. In our model for matter particles they are electromagnetic waves spinning in a convoluted ring as a standing wave oscillating at a specific frequency. Let us examine the neutrons being squashed in this shrinking core.

The size of each neutron depends on its De Broglie wavelength:

$$\lambda_{b} = h/mc$$
$$= 6.626 \times 10^{-34}\text{Js}/(1.675 \times 10^{-27}\text{kg} \times 2.998 \times 10^{8}\text{m/s})$$
$$\lambda_{b} = 1.3195 \times 10^{-15}\text{m}$$

The radius of the neutron as a circular standing wave where the circumference equals the De Broglie wavelength equals:

$$r_{n} = 1.3195 \times 10^{-15}\text{m}/2\pi = 2.1 \times 10^{-16}\text{m}$$

What is interesting about the standing wave nature of particles is that the De Broglie wavelength which determines the particle size, is inversely proportional to its mass, the more massive the particle the smaller its size. This is why the nucleons in a nucleus are much smaller than the surrounding electrons. At the limit between the neutron star and a black hole the standing wave neutrons in the core are being squashed. So what happens to the wave if the collapse continues? Are the electromagnetic standing waves of the neutrons squashed to a point? The answer to this question may lie in the quantum nature of particles and waves. Standing waves which make up particles can only oscillate at fixed integer numbers (n) of whole wavelengths:

$$\lambda_b = n \times h/mc$$

A fundamental particle will oscillate at its fixed frequency/wavelength where n = 1 whole wavelength and in a specific spinning convoluted form where charge and mass are expressed. The neutron has three loops to its electromagnetic spin (which we see as quarks) and compressing this standing wave will shorten its wavelength and increase the oscillation frequency – hence simultaneously reducing the neutron's size and increasing its mass. A shorter De Broglie wavelength means a smaller more massive particle. Therefore the neutrons in our shrinking black hole are becoming smaller and more massive! Neutrons, therefore may be squashed into heavy particles, hadrons, but will this process continue with the heavy neutrons becoming increasingly more massive and smaller in size until infinite density?

If a sound standing wave is set up between two metal plates using a signal generator and a small loud speaker at a fixed audible frequency where the distance between the plates is exactly half a wavelength (fundamental frequency) and the plates are quickly moved together the wave being squashed between the plates reflects back and forth adopting a shorter wavelength standing wave - a higher audible frequency.

This can be heard above the fixed frequency generated by the loudspeaker. If the plates are moved slowly together there is no audible shift in pitch. This increase in pitch is similar to a Doppler effect due to the moving sound source – the reflecting plate, when the plates are moved apart quickly the pitch drops as the wavelength increases.

It seems therefore that squashed standing waves produce increase in frequency and therefore mass, but is this process unlimited? For a 'free' standing wave it is because by the process of interference of the reflected waves the wave adopts a half wavelength which is exactly equal to the distance between the plates. In quantum mechanics particles have specific quantities: the particle standing wave adopts a fixed size (whole De Broglie wavelength), spin, mass and electromagnetic convolution – mass particles can only exist in fixed quantum mass-energies. Therefore according to the principles of quantum mechanics as the neutrons get squashed in the core beyond the Schwarzschild radius there may be a new of barrier preventing total collapse – **primary hadron degeneracy pressure.**

If this is the case the neutron star (core) squashed beyond the Schwarzschild radius will stop contracting when the neutrons are squashed to the next quantum jump of mass-energy particle – it will become a super dense heavy hadron star of fixed size and mass, but will appear as a black hole because it gravity is strong enough to prevent light escaping. If the mass of the collapsing core is greater than 2.5 solar masses then there may be sufficient gravitational collapse to squeeze the heavy hadrons to the next mass-energy quantum jump of fixed mass and size – **secondary hadron degeneracy pressure.**

For larger mass stars this process may continue producing smaller but fixed size hadron stars all which appear as black holes. The hadron core will not collapse to a singularity – **quantum hadron degeneracy** pressure prevents this even for the largest mass stars of around 100 solar masses with a core mass of around 30 solar masses. There is therefore possibly a fixed range of black holes, hadron stars, in our universe arising from the death of main sequence stars. Concerning these black holes arising out of the death of a main sequence star and the black holes found in the centres of galaxies, these hadron stars may be *layered* with *hadron particles* of increasing quantum mass and decreasing quantum size.

2. Neutron Density of a Minimum Mass Black Hole before Further Collapse

In Our minimum mass black hole of 2.5 solar masses we have stated that just before the shrinking core reaches the Schwarzschild radius (inside of which it is a black hole) the super dense core is composed of neutrons, to shrink under the forces of gravity beyond this limit after which light itself cannot exist gravity must overcome neutron degeneracy pressure. Neutrons are hadrons and are subject to the **strong nuclear force**, which like protons in a nucleus are held together by this attractive force, but at very short distances this force become repulsive.

$$\text{Mass of neutron} = 1.675 \times 10^{-27} \text{kg}$$

$$\text{De Broglie wavelength size of neutron } \lambda b = 1.3195 \times 10^{-15} \text{m}$$

$$\text{Radius of standing wave ring } r_n = 2.1 \times 10^{-16} \text{m}$$

$$\text{Nuclear density} = 1.72 \times 10^{17} \text{kg/m}^3$$

$$\text{Nuclear radius} = 1.2 \times 10^{-15} \text{m}$$

Given that the density of a 2.5 solar mass core at the Schwarzschild radius equals 2.946×10^{18} kg/m^3, how closely packed would these neutrons be? Since nuclear density is 1.72×10^{17} kg/m^3, then as said before the neutron star is 17.13 times denser, this means that the nucleons are on average 17 times closer than nucleons in a nucleus. <u>At this range the strong nuclear force is repulsive.</u> At this density and spacing neutron degeneracy pressure is overcome, but instead of the neutron core shrinking to a black hole singularity the black hole will make a quantum jump to a size which is less than the Schwarzschild radius, but nevertheless fixed by the size of the newly formed heavier hadrons, if we assume at this point a quantum jump of double the neutrons particle frequency the collapse will halt at exactly half the size of the Schwarzschild radius of 3.7km! So assuming the largest mass star is 100 solar masses with a Schwarzschild radius of 46.34km, and then a similar quantum jump in the mass-energy of the neutrons, it terms of frequency one octave higher, will give the biggest black hole hadron star with a radius of 23.17km.

In a simple quantum view the range of sizes for black hole hadron stars is:

3.7 – 23.2 km

Definitely not singularities!

Please note: It must be remembered that in reality heavy hadrons do not have masses that are simple multiples of the mass of the neutron, although they oscillate at higher frequencies (particle mass is proportional to frequency) they are composed of different configurations of quarks (standing wave loops) which have themselves different masses and mode of oscillation (convolution). We can think of these increasing mass hadrons like harmonic notes along an octave which vary in a specific pattern between a fundamental note (f = 1) and an octave higher (f = 2). Therefore the actual size of these quantum particle hadron stars will have a mass and radius which depends on the type of hadrons being formed and as we have said a single hadron star may very well have shells of increasing mass hadrons towards the centre of the hadron black hole.

3. Light Speed and the Space Vacuum at the Event Horizon

In our model of gravity one can easily imagine the tremendous power of the accelerating flux of space vacuum energy near a black hole, anything approaching the black hole will be sucked along with this flux of increasing energy density.

Remember our model for gravity is explained by two variables:

1. An accelerating flux of space vacuum energy into a mass.

2. An increasing energy density as a result of this inward flux.

Each follows and inverse square law and both aspects of this space vacuum energy flux into the mass produce what we see as a gravitational field or space-time curvature. This model explains the gravitational acceleration of a mass, length contraction, time dilation and all the other affects associated with Einstein's general relativity theory. At the event horizon the gravitational curvature of space-time is so great that not even light can escape – can our model of gravity explain this phenomenon?

Let us imagine riding a beam of light and trying to escape at the event horizon. If we could see the accelerating rush of space vacuum energy we would feel like we are rowing a boat against a mighty flux of water. Now let us assume we can equal the acceleration of the energy flux at each stroke, would we escape? The answer of course is no, because for between each stroke we would have a small time delay before we put the oars back in the water. During this small time we would loose ground. Let us compensate and row with a slightly greater accelerating force on each stroke – now we would keep up but still get nowhere, just like light at the event horizon. In our model of gravity the time delay for the light wave is caused by absorption and re-emission of light photons by the space vacuum mass-energy particles – the ether which restricts the speed of light and at the event horizon of a black hole this mass-energy density is extremely high and increases as you fall closer. Inside the event horizon the speed of light because of this time delay through the ether is insufficient for light to escape.

4. Hawking Radiation and Black Holes

Even though light cannot escape a black hole the physicist Steven Hawking predicted that black holes should emit radiation, what has been called since Hawking radiation. This he explains by examining the fate of virtual particles emerging from the space vacuum zero point energy. Since these virtual particles are created in pairs: matter and antimatter particles, three things could happen to them:

1. Both particles are pulled into the black hole

2. Both particles escape the black hole

3. One particle escapes while the other is pulled into the black hole

In the third case the particle which escapes can become real, by absorbing the energy of a gamma photon; this particle will quickly annihilate its matter/antimatter opposite and emit a gamma photon which escapes the black hole as Hawking radiation, so in theory black holes should emit radiation from the event horizon. The problem is Hawking radiation has never been detected. The reason why I believe is that in the conventional model of gravity it assumes that the space-time matrix which produces the virtual pairs of particles is stationary even at the edge of a black hole where space-time is extremely distorted. In our model of gravity as a space vacuum energy flux these pairs of virtual particles rapidly accelerate into the black hole past the event horizon and are therefore BOTH captivated by the black hole by the extreme flux before any emission can occur.

5. Einstein's Equivalence Principle

'All objects fall at the same rate independent of the type of substance'

According to Einstein no matter what substance an object is made of it will accelerate downwards at the same rate as any other substance i.e.

'g' is the same at a given point on the Earth's surface for any type of material. This has been proven to extreme accuracy, but physicists today are trying to test this principle to the limit – to the atomic scale. Experiments are under way to see if atoms of different isotopes of rubidium show any difference in how gravity affects them. If we consider gravity as not a force and what we see as a curvature of space-time is in fact a flux of the space-vacuum energy into a mass producing a variation in its density then 'g' would indeed be a constant for all types of matter. Mass accelerates with the accelerating flux of energy. Considering whether there is a small variation in 'g' between different types of atoms is a possibility since different atoms have different electron configurations, remembering it is the spin of the electron, its angular momentum which we see as mass (and related to the constant h) which causes the gravitational space vacuum energy flux.

There is the possibility that only electron spin in matter produces this gravitational flux of energy and that the presence of other particles such as protons and neutrons in the nucleus of atoms slightly modify this flux rate. The other possibility is that all fundamental particles since they have spin produce gravitational energy flux. The latter is more likely to be true since 'g' is proportional to mass. Looking back at the equation:

g is proportional to Mass which is proportional to **hfu_0e_0**

Then the higher the mass of the particle the higher the oscillation frequency of the particle wave, the shorter its wavelength and the smaller the radius of the standing wave particle, since all the other terms are constants except (f) then increasing f must produce an increase in what we see as mass and therefore increasing f also increases the gravitational energy flux rate, this makes sense since the particle vortex has a higher frequency and therefore greater absorption of space vacuum energy therefore producing a greater flux of energy. If considering the space vacuum energy pressure difference between the inside and outside of the particle standing wave, by the Casimer effect, the smaller the particle the greater the pressure difference and therefore the greater the flux rate. In either case increasing the mass of a particle increases the standing particle wave flux effect on the space vacuum energy and therefore there is an increase in 'g.'

6. Unification of Field Theory and Quantum Mechanics

One 'thorn in the flesh' of modern physics is unifying Einstein's field theory with quantum mechanics. In Einstein's world, the curvature of space-time is smooth no matter at what level one looks at it - space-time curvature is a continuously varying matrix, even if you looked at it on the smallest sub atomic level. This does not match modern quantum mechanics where the mass-energy of particles is not continuously varying, in this world particles and photons of light have lumpiness i.e. are quantized by specific amounts with no possible values in between. In the world of quantum mechanics particles can only shift mass-energy levels in steps and have probabilities of being at one energy level or another. So shouldn't gravity on the smallest level be lumpy, not according to Einstein 'God does not play dice with the universe.' How can these two worlds, Einstein's smooth gravity field and quantum mechanics, be unified? This has been a modern dilemma for physicists and has we have seen this may be answered by the flux mechanism of gravity model. In our model of gravity Einstein's field theory and modern quantum mechanics are right, it depends which way you look at space-time curvature – the gravity field.

1. A gravitational accelerating flux of energy cannot have lumpiness, it is not particulate it is simply an increasing velocity of the energy flux into a mass which we see as a smooth curvature of space-time.

2. On the other hand looking at the gravitational curvature of space-time around a mass from an increase in energy particle density point of view, which remember is particulate in nature, then the gravitational field is lumpy, but the increasing energy density is an average at of all the seething quantum foam.

Quantum mechanics stands and so does Einstein's view that gravitational space-time curvature is smoothly varying.

QED!

7. Inflationary Period

The inflationary period is the term used in cosmology to describe the brief time in the very early universe when, according to inflation theory, the universe was expanding exponentially. It is believed to have occurred between 10^{-36} and 10^{-32} seconds, after the Big Bang when the temperature of the universe was 10^{32} to 10^{27} Kelvin. During this time, the size of the universe increased by a factor of 10^{50} from an initial size of 10^{-26} meters in diameter (a hundred billion times smaller than a proton) to approximately one hundred million light years (10^{24} m) in diameter. This exponential inflation solves many of the problems of a simple big bang, such as the flatness of the universe.

Rapid expansion, during the inflationary era between 10^{-36} seconds and 10^{-32}, happened because during this time there were no gravity particles, the temperatures were too high for matter particles to form; only light photons existed in this epoch which are not vortices and therefore have no gravity. The immediate space vacuum energy density exerted an enormous pressure on the volume of space-time and thus caused a rapid expansion. As the universe expanded and cooled the temperatures became low enough for ordinary fundamental gravity particles to form, electrons, protons and quarks, *gravity now emerges as a force* in the universe and slows down the expansion – in other words the standing waves vortices started gravitational flux and immediately put the brakes on the expansion. This happened at 10^{-34} seconds. In modern theoretical physics this was about the time when gravity as the first of the four forces in nature emerged due to symmetry breaking – where fluctuations on a quantum level reach a critical point. Initially the space vacuum energy density was extremely high creating high space vacuum pressure resulting in strong gravity. With time, by expansion and the settling down of early massive pulsars the energy density of the universe was reduced and gravity became weaker, this we see as an accelerating expansion of the universe.

8. Entanglement

Maybe entanglement operates at finite speeds, but so fast we cannot measure its operating speed! Let me explain this reasoning further. We know that light travels as an electromagnetic transverse wave through the space vacuum and that the electric and magnetic constants of the ether, which are related to absorption and emission times by the ether virtual particles, limit the speed of light, so by what mechanism does entanglement appear to bypass this limiting space vacuum energy density?

Longitudinal compression waves through a medium travel very much faster than transverse waves through the same medium. Now if we consider that light is a transverse electromagnetic wave travelling though the space vacuum energy density of virtual particles at a speed of c then a compression wave through the same nuclear dense material would travel very much faster than the speed of light! But doesn't this violate Einstein's principle that nothing can travel faster than the speed of light? But remember there is no magical value about the speed of light it is simply the speed determined by the space vacuum energy density where photons of light take a finite time to be absorbed and re-transmitted by the virtual particles. Maybe, therefore compression waves through the space-vacuum travel many times faster than light speed and would therefore to us and the limits of our measurement appear instantaneous. Maybe there is a particle information exchange in the universe which is by another mechanism other than electromagnetic photons exchange, perhaps what we see as instantaneous entanglement of particles is not instantaneous at all, perhaps its just that the effects are too fast too measure.

9. Energy Density Engineering

Reduce the energy density of the space vacuum to reduce the drag on a mass, therefore reducing the force needed to accelerate a mass (spaceship) or produce a density gradient to accelerate a mass or alternatively produce an accelerated flux of the space vacuum to create a gravitational field.

There are many far reaching implications to understanding a mechanism for gravity, maybe the model of energy flux discussed in this book comes close to a true explanation of the way gravity works, or maybe it just points the way to a deeper understanding of the most elusive force of the universe. This journey started with the adage 'a watched kettle never boils,' and maybe, in time, we will know why.

CHAPTER 15

Reflections and Unequivocal Evidence

Postulates of the Quantum Gravitational Flux Theory (QGFT) of Gravity

The quantum gravitational flux theory model has the following distinctive features:

1. Fundamental particles are standing wave electromagnetic vortices causing an accelerating flux of the space vacuum energy density into a mass particle.

2. This accelerating energy flux is aided and maintained by the space vacuum energy density pressure of the universe, because of a high energy density outside the particle vortex and a low near zero energy density inside the particle vortex (Casmir Effect).

3. Absorption of the space vacuum energy density virtual particles by the particle vortex maintains the particle's vortex spin at an equilibrium ground state and each absorption is seen as a shift in the energy/wavelength/frequency of the standing wave (Lamb Shift).

4. This space vacuum accelerating energy flux into a sink mass causes matter particles, which also have a similar flux of space-time into their rest mass, to accelerate towards each other.

5. A mass caught up in this flux appears weightless and in free fall – it is at rest with the gravitational flux.

6. When a mass particle comes to rest on the surface of the gravitating mass it experiences a drag force from the accelerating space-vacuum through it, this is seen as its weight.

7. A mass particle in constant motion through the space vacuum energy density does not experience a drag force from the energy density because it is in electromagnetic energy equilibrium with it.

8. A mass particle accelerating through the space vacuum experiences a drag force because it is no longer in a state of electromagnetic energy

equilibrium with the space vacuum, therefore it requires a Newton force of F=ma, where 'a' is the acceleration of the mass, to accelerate through the space vacuum. The greater acceleration through the space vacuum or the greater the mass in contact with the space vacuum the greater the drag force. Newton's first and second laws of motion are proof that the acceleration of a mass in a gravitational field must likewise be due to the fact that the space-vacuum itself must be accelerating through the mass. The mass must be in an unequal state of electromagnetic energy equilibrium with the accelerating space vacuum and therefore experiences a drag force where w =mg, where 'g' is the acceleration of the space vacuum.

This postulate may be the cornerstone of the QGFT: There is no distinction between the two: a mass accelerating through a space vacuum or the space vacuum accelerating through a mass – each produces a drag force.

9. The decreasing volume of the accelerating space vacuum energy into a sink mass produces an increase in the energy density as it approaches the mass. This increasing energy density produces all the known General Relativity effects of a gravitational field, including a reduction in the speed of light, space and time dilation, refraction and slowing down of light (Shapiro Delay)

10. The accelerating flux follows and inverse square law which is seen as a radial gravitational field.

11. In quantum gravitational flux theory (QGFT) light does not have gravity, only mass particles produce a gravitational field. A light photon is a propagating free wave cannot cause a gravitational flux of this medium, only true particles with rest mass which are standing wave vortexes can produce a gravitational flux of the space vacuum. This has serious implications for the very early universe and expansion.

12. The quantum gravitational flux theory unifies quantum mechanics and the smooth field variations of Einstein's General Relativity: A gravitational flux is a smooth continuously varying flux of the space vacuum, but is composed of a quantized particulate energy density of virtual and real particles.

13. The fine structure constant (α), sometimes known as the God constant, which determines the strength of electromagnetic interactions enables the Lamb Shift of electrons immersed in the space vacuum see of virtual particles and photons to be calculated and gives rise to the

magnitude of G and may be a quantum explanation and derivation of G and gravitational force in the formulae:

$$\Delta E_L = \alpha^5 m_e c^2 \times k_L$$

$$G_L = \Delta E_L / E_e = \alpha^5 \times k_L$$

$$\textbf{Gravitational Force} = -\alpha^5 k_L Mm/r^2$$

14. The base units of G derived from Newton's Universal Gravity formula are equal to:

Rate of change of volume per unit of gravitating mass

Inferring that G determines the magnitude of the rate of change of a collapsing sphere of the space vacuum around a mass, thus suggesting that gravity is an accelerating influx of the space vacuum's energy density. The units of G may therefore be another cornerstone for QGFT.

A Final Word from the Author

'It's humbling to think that after some four hundred years, Newton's laws of motion and gravity give us the most important clues as to the true nature of gravity, even in our world of Einstein's Special and General Relativity theories and modern quantum physics.'

Albert Einstein Quotes

Imagination is more important than knowledge.

The only real valuable thing is intuition.

Anyone who has never made a mistake has never tried anything new.

Everything should be made as simple as possible, but not simpler.

The only thing that interferes with my learning is my education.

"The most incomprehensible thing about the world is that it is comprehensible.

We can't solve problems by using the same kind of thinking we used when we created them.

The important thing is not to stop questioning.
Curiosity has its own reason for existing.

Index